Hilbert Modular Forms: mod p and p-Adic Aspects

MEMOIRS
of the
American Mathematical Society

Number 819

Hilbert Modular Forms: mod p and p-Adic Aspects

F. Andreatta
E. Z. Goren

January 2005 • Volume 173 • Number 819 (fourth of 5 numbers) • ISSN 0065-9266

American Mathematical Society
Providence, Rhode Island

2000 *Mathematics Subject Classification.*
Primary 11G18, 14G35, 11F33, 11F41.

Library of Congress Cataloging-in-Publication Data

Andreatta, F. (Fabrizio), 1972–
Hilbert modular forms : mod p and p-adic aspects / F. Andreatta, E. Z. Goren.
p. cm. — (Memoirs of the American Mathematical Society, ISSN 0065-9266 ; no. 819)
"Volume 173, number 819 (fourth of 5 numbers)."
Includes bibliographical references.
ISBN 0-8218-3609-9 (alk. paper)
1. Arithmetical algebraic geometry. 2. Hilbert modular surfaces. 3. Forms, Modular. 4. Moduli theory. I. Goren, Eyal Z. (Eyal Zvi), 1963– II. Title. III. Series.
QA3.A57 no. 819
[QA242.5]
510 s—dc22
[516.3′5] 2004057485

Memoirs of the American Mathematical Society

This journal is devoted entirely to research in pure and applied mathematics.

Subscription information. The 2005 subscription begins with volume 173 and consists of six mailings, each containing one or more numbers. Subscription prices for 2005 are \$606 list, \$485 institutional member. A late charge of 10% of the subscription price will be imposed on orders received from nonmembers after January 1 of the subscription year. Subscribers outside the United States and India must pay a postage surcharge of \$31; subscribers in India must pay a postage surcharge of \$43. Expedited delivery to destinations in North America \$35; elsewhere \$130. Each number may be ordered separately; *please specify number* when ordering an individual number. For prices and titles of recently released numbers, see the New Publications sections of the *Notices of the American Mathematical Society.*

Back number information. For back issues see the *AMS Catalog of Publications.*

Subscriptions and orders should be addressed to the American Mathematical Society, P. O. Box 845904, Boston, MA 02284-5904, USA. *All orders must be accompanied by payment.* Other correspondence should be addressed to 201 Charles Street, Providence, RI 02904-2294, USA.

Memoirs of the American Mathematical Society is published bimonthly (each volume consisting usually of more than one number) by the American Mathematical Society at 201 Charles Street, Providence, RI 02904-2294, USA. Periodicals postage paid at Providence, RI. Postmaster: Send address changes to Memoirs, American Mathematical Society, 201 Charles Street, Providence, RI 02904-2294, USA.

This publication is indexed in *Science Citation Index®*, *SciSearch®*, *Research Alert®*, *CompuMath Citation Index®*, *Current Contents®/Physical, Chemical & Earth Sciences.*
Printed in the United States of America.

∞ The paper used in this book is acid-free and falls within the guidelines established to ensure permanence and durability.
Visit the AMS home page at `http://www.ams.org/`

10 9 8 7 6 5 4 3 2 1   10 09 08 07 06 05

Contents.

Abstract.

We study Hilbert modular forms in characteristic p and over p-adic rings. In the characteristic p theory we describe the kernel and image of the q-expansion map and prove the existence of filtration for Hilbert modular forms; we define operators U, V and Θ_χ and study the variation of the filtration under these operators. Our methods are geometric – comparing holomorphic Hilbert modular forms with rational functions on a moduli scheme with level-p structure, whose poles are supported on the non-ordinary locus.

In the p-adic theory we study congruences between Hilbert modular forms. This applies to the study of congruences between special values of zeta functions of totally real fields. It also allows us to define p-adic Hilbert modular forms "à la Serre" as p-adic uniform limit of classical modular forms, and compare them with p-adic modular forms "à la Katz" that are regular functions on a certain formal moduli scheme. We show that the two notions agree for cusp forms and for a suitable class of weights containing all the classical ones. We extend the operators V and Θ_χ to the p-adic setting.

2000 *Mathematics Subject Classification.* Primary 11G18, 14G35, 11F33, 11F41.

Key words: Congruence, Hilbert modular form, Hilbert modular variety, zeta function, filtration.

1 Introduction.

This paper is concerned with developing the theory of Hilbert modular forms along the lines of the theory of elliptic modular forms. Our main interests in this paper are:

(i) to determine the ideal of congruences between Hilbert modular forms in characteristic p and to find conditions on the existence of congruences over artinian local rings. This allows us to derive explicit congruences between special values of zeta functions of totally real fields, to establish the existence of filtration for Hilbert modular forms, to establish the existence of p-adic weight for p-adic modular forms (defined as p-adic uniform limit of classical modular forms) and more;

(ii) to construct operators U, V, Θ_ψ (one for each suitable weight ψ) on modular forms in characteristic p and to study the variation of the filtration under these operators. This allows us to prove that every ordinary form has filtration bounded from above;

(iii) to show that there are well defined notions of a Serre p-adic modular form and of a Katz p-adic modular form and to show that the two notions agree for a suitable class of weights containing all the classical ones. Our argument involves showing that every q-expansion of a mod p modular form lifts to a q-expansion of a characteristic zero modular form. We extend the theta operators Θ_ψ to the p-adic setting – their Galois theoretic interpretation is that of twisting a representation by a Hecke character.

Our approach to modular forms is emphatically geometric. Our goal is to develop systematically the geometric and arithmetic aspects of Hilbert modular varieties and to apply them to modular forms. As to be expected in such a project, we use extensively the ideas of N. Katz [Ka1], [Ka2], [Ka3], [Ka4] and J.-P. Serre [Se], of the founders of the theory in the case of elliptic modular forms, and we have benefited much from B. Gross' paper [Gr]. In regard to previous work on the subject, we mention that some of the constructions and methods in this paper were introduced by the second named author, in the unramified case, in [Go1], [Go2], and the congruences we list for zeta functions may be derived from the work of P. Deligne and K. Ribet [DeRi]. For this reason we restrict our discussion to zeta functions, though the same reasoning applies to a wide class of L-functions.

We now describe in more detail the main results of this paper.

Let L be a totally real field of degree g over $\mathbf{Q}$. Let K be a normal closure of L. Let $N \geq 4$ be an integer prime to p. Let $\mathfrak{M}(S, \mu_N)$ be the fine moduli scheme parameterizing polarized abelian schemes over S with RM by O_L and μ_N-level structure; see 3.2. A Hilbert modular form defined over an O_K-scheme S has a weight $\psi \in \mathbb{X}_S$, where $\mathbb{X}_S$ is the group of characters of the algebraic group $\mathcal{G}_S = \mathrm{Res}_{O_L/\mathbf{Z}}\mathbf{G}_{m,O_L} \times_{\mathrm{Spec}(\mathbf{Z})} S$. We shall mostly be concerned with weights obtained from the characters $\mathbb{X}$ of $\mathcal{G}_{O_K} = \mathrm{Res}_{O_L/\mathbf{Z}}\mathbf{G}_{m,O_L} \times_{\mathrm{Spec}(\mathbf{Z})} \mathrm{Spec}(O_K)$; we shall use the notation $\mathbb{X}_S^U$ ("U" for universal) to the denote the group of characters of $\mathcal{G}_S$ induced from $\mathbb{X}$ by base change. See 4.1.

The group $\mathbb{X}$ is a free abelian group of rank g and has a positive cone $\mathbb{X}^\dagger$ generated by the characters coming from the embeddings $\sigma_1, \dots, \sigma_g \colon L \to K$. Indeed, the

Received by the editor August 1, 2001.

map $O_L \otimes_{\mathbf{Z}} K \cong \oplus_{i=1}^{g} K$ induces a splitting of the torus $\mathcal{G}_K$, and hence canonical generators of $\mathbb{X}$ that we denote accordingly by $\chi_1, \dots, \chi_g$, and call the *basic characters*. A complex Hilbert modular form of weight $\chi_1^{a_1} \dots \chi_g^{a_g}$ is of weight $(a_1, \dots, a_g)$ in classical terminology.

It is important to note that $\mathbb{X}_S^U$ depends very much on S. For example, assume that L is Galois and $S = \mathrm{Spec}(O_L/p)$, with p an inert prime in L, then $\mathbb{X} \cong \mathbb{X}_S^U$, while if p is totally ramified in L, $p = \mathfrak{P}^g$, then $\mathbb{X}_S^U \cong \mathbf{Z}$; in this case, letting Ψ denote the reduction of any basic character χ_i, we obtain that an O_L-integral Hilbert modular form of weight $\chi_1^{a_1} \dots \chi_g^{a_g}$ reduces modulo p to a modular form of weight $\Psi^{a_1 + \dots + a_g}$.

We denote the Hilbert modular forms defined over S, of level μ_N and weight χ by $\mathbf{M}(S, \mu_N, \chi)$.

Let p be a rational prime. Let k be a finite field of characteristic p, which is an O_K-algebra. Let $\mathbb{X}_k(1)$ be the subgroup of $\mathbb{X}_k$ consisting of characters χ that are trivial on $\left(O_L/(p)\right)^*$ under the map $\left(O_L/(p)\right)^* \hookrightarrow \mathcal{G}_k(k) = \left(O_L \otimes_{\mathbf{Z}} k\right)^* \xrightarrow{\chi} \mathbf{G}_m(k) = k^*$. It is proven in 4.4 that the map $\mathbb{X} \to \mathbb{X}_k$ is surjective and, in particular, $\mathbb{X}_k = \mathbb{X}_k^U$. This allows us to define a positive cone $\mathbb{X}_k^+$ in $\mathbb{X}_k$ as follows. For every i there exists $1 \leq \tau(i) \leq g$ such that the image of $\chi_i^p \chi_{\tau(i)}^{-1}$ in $\mathbb{X}_k^U$ is in $\mathbb{X}_k(1)$. The character $\chi_i^p \chi_{\tau(i)}^{-1}$ in $\mathbb{X}_k^U$ does not depend on the choice of $\tau(i)$. The positive cone in $\mathbb{X}_k$ is the one induced by these generators. The positive cone induces an order $\leq_k$ on $\mathbb{X}_k$; we say that $\tau_1 \leq_k \tau_2$ if $\tau_1^{-1} \tau_2$ belongs to the positive cone. Note that we have provided $\mathbb{X}_k(1)^+ := \mathbb{X}_k(1) \cap \mathbb{X}_k^+$ with a canonical set of generators.

For every character $\psi \in \mathbb{X}_k(1)^+$ we construct in 7.12 a holomorphic modular form h_ψ over k. By 7.14 it has the property that its q-expansion at any $\mathbf{F}_p$-rational cusp is 1. Moreover, the ideal $\mathcal{I}$ of congruences

$$\mathcal{I} := \mathrm{Ker}\left\{ \bigoplus_{\chi \in \mathbb{X}_k} \mathbf{M}(k, \mu_N, \chi) \xrightarrow{q\text{-exp}} k[\![q^\nu]\!]_{\nu \in M} \right\}$$

(where M is a suitable O_L-module depending on the cusp used to get the q-expansion) is given by

$$\left(h_\psi - 1 : \psi \in \mathbb{X}_k(1)^+\right).$$

It is a finitely generated ideal and a canonical set of generators is obtained by letting ψ range over the generators for $\mathbb{X}_k(1)^+$ specified above; see 7.22; Cf. [Go2].

Again, it may beneficial to provide two examples. Assume that L is Galois. If p is inert in L, we may order $\sigma_1, \dots, \sigma_g$ cyclically with respect to Frobenius: $\sigma \circ \sigma_i = \sigma_{i+1}$. Let $k = O_L/(p)$. Then $\mathbb{X}_k(1)$ and $\mathbb{X}_k(1)^+$ are generated by the characters $\chi_1^p \chi_2^{-1}, \dots, \chi_i^p \chi_{i+1}^{-1}, \dots, \chi_g^p \chi_1^{-1}$. Note that this positive cone is different from the one obtained from $\mathbb{X}^\dagger$ via the reduction map. The kernel of the q-expansion map is generated by g relations $h_1 - 1, \dots, h_g - 1$, where $h_i = h_{\chi_{i-1}^p \chi_i^{-1}}$ is a modular form of weight $\chi_{i-1}^p \chi_i^{-1}$. On the other hand, when $p = \mathfrak{P}^g$ is totally ramified, $k = O_L/\mathfrak{P}$, we find that $\mathbb{X}_k(1)$ is generated by the characters $\chi_1^{p-1}, \dots, \chi_g^{p-1}$ that are all the same character Ψ^{p-1} in $\mathbb{X}_k^U$ and the q-expansion kernel is generated by a single relation $h_{\Psi^{p-1}} - 1$, where $h_{\Psi^{p-1}}$ is a modular form of weight Ψ^{p-1}.

We offer two constructions of the modular forms h_ψ; see 7.12 and 7.18. One construction allows us to prove in 8.18 that the divisor of h_ψ, for ψ one of the canonical generators of $\mathbb{X}_k(1)^+$, is a reduced divisor. The other construction is related to a compactification of $\mathfrak{M}(k, \mu_{Np})$.

The proof of the theorem on the ideal of congruences is based on the isomorphism between the ring $\oplus_{\chi \in \mathbb{X}_k} \mathbf{M}(k, \mu_N, \chi)/\mathcal{I}$ of *modular forms as q-expansions* and the ring of regular *functions* on the quasi-affine scheme $\mathfrak{M}(k, \mu_{Np})$. The latter scheme can be compactified by adding suitable roots of certain of the sections h_ψ. This isomorphism creates a *dictionary* between modular forms of level μ_N and meromorphic functions on the compactification $\overline{\mathfrak{M}}(k, \mu_{Np})$ that are regular on $\mathfrak{M}(k, \mu_{Np})$. Under this dictionary, the weight of a modular form, a character in $\mathbb{X}_k$, is mapped to an element of $\mathbb{X}_k/\mathbb{X}_k(1)$, the k^*-valued characters of the Galois group $\left(O_L/(p)\right)^*$ of the cover $\mathfrak{M}(k, \mu_{Np}) \to \mathfrak{M}(k, \mu_N)^{\mathrm{ord}}$. The exact behavior of the poles of such a meromorphic function is related to a minimal weight (with respect to the order $\leq_k$ on $\mathbb{X}_k$), called *filtration*, from which a q-expansion may arise. Here the explicit description of the compactification is invaluable in studying the properties of the filtration.

This dictionary also allows us to define operators that clearly depend only on q-expansions, like U, V and suitable Θ_ψ operators, first as operators on functions on $\mathfrak{M}(k, \mu_{Np})$, and then as operators on modular forms (see §§ 12-14). This enables us to read some of the finer properties of these operators from the corresponding properties on $\mathfrak{M}(k, \mu_{Np})$. Our main results are the following:

- There exists a notion of filtration for Hilbert modular forms: a q-expansion arising from a modular form f arises from a modular form of minimal weight $\Phi(f)$ with respect to $\leq_k$; see 8.20.
- There exists a linear operator $V: \mathbf{M}(k, \chi, \mu_N) \to \mathbf{M}(k, \chi^{(p)}, \mu_N)$, whose effect on q-expansions is $\sum a_\nu q^\nu \mapsto \sum a_\nu q^{p\nu}$. The character $\chi^{(p)}$ is the character induced from χ by composing with Frobenius. (Concretely, in the inert case $\chi_i^{(p)} = \chi_{i-1}^p$, and in the totally ramified case $\Psi^{(p)} = \Psi^p$.) We have $\Phi(Vf) = \Phi(f)^{(p)}$. The operator V comes from the Frobenius morphism of $\mathfrak{M}(\mathbf{F}_p, \mu_{Np})$. See 13.1, 13.5, 13.9 and 15.12.
- For every basic character ψ there exists a k-derivation Θ_ψ taking modular forms of weight χ to modular forms of weight $\chi\psi^{(p)}\psi$. Its effect on q-expansions at a suitable cusp is given by $\sum a(\nu) q^\nu \mapsto \sum \psi(\nu) a(\nu) q^\nu$. See 12.38–12.40. The behavior of filtration under such operator involves too much notation to include here and we refer the reader to 15.10. In the inert case, $\Phi(f) = \chi_1^{a_1} \cdots \chi_g^{a_g}$, we have $\Phi(\Theta_{\chi_i} f) \leq_k \Phi(f) \chi_{i-1}^p \chi_i$, and $\Phi(\Theta_{\chi_i} f) \leq_k \Phi(f) \chi_i^2$ iff $p | a_i$ (note: $\chi_i^2 = (\chi_{i-1}^p \chi_i)/(\chi_{i-1}^p \chi_i^{-1})$). If p is completely ramified the result resembles the elliptic case: if $\Phi(f) = \Psi^a$ then $\Phi(\Theta_\Psi f) \leq_k \Psi^{a+p+1}$ and $\Phi(\Theta_\Psi f) \leq_k \Psi^{a+2}$ iff $p|a$.
- There exists a linear operator U taking holomorphic modular forms of weight χ to meromorphic modular forms of weight χ. See 14.7. The effect of U on q-expansions is

$$\sum a_\nu q^\nu \mapsto \sum a_{p\nu} q^\nu.$$

If χ is "positive enough" (see 14.7 for a precise statement) then U takes holomorphic modular forms of weight χ to holomorphic modular forms of weight χ. Every ordinary modular form has the same q-expansion as an ordinary modular form whose weight is bounded from above (see 15.15). For example, such weight lies in $(-\infty, p+1]^g$ for p inert. It is $\leq g(p+1)$ if p is totally ramified.

We always have an identity of q-expansions between $V \circ U$ and a certain operator constructed from the operators Θ_{χ_i}; see 14.7.1 for the general formula.

For instance, if p is totally split then

$$VUf(q) = \prod_{i=1}^{g}\big(\mathrm{Id} - \Theta_{\chi_i}^{p-1}\big)f(q),$$

while if p is inert we have

$$VUf(q) = \big(\mathrm{Id} - \prod_{i=1}^{g}\Theta_{\chi_i}^{p-1}\big)f(q).$$

The product appearing in the split case is typical of the general case. If we write the operator on the right hand side as $\mathrm{Id} - \Lambda$, then on the one hand we have the classical identity $V \circ U = \mathrm{Id} - \Lambda$, but on the other hand Λ has a complicated expression (involving addition and composition) in terms of the basic theta operators which makes its study difficult. However, in the inert case we can improve and further our results.

- Assume that p is inert in L. In this case we study in §16 two phenomena. The first is modular forms of parallel weight, the notion of *parallel filtration* (the definition can be made in the general case as well) and the behavior of those under U, V and $\theta = \Theta_{\chi_1}\cdots\Theta_{\chi_g}$. For example, in 16.13 we prove that every ordinary form of parallel weight is congruent to an ordinary form of parallel weight lying in the range $[2, p+1]$.

  The second phenomenon is that of Θ_{χ_1}-cycles. This suffices to study all q-expansions obtained by "twists". Following N. Jochnowitz we establish several combinatorial facts concerning such cycles. For $g > 1$ it appears that the structure is much richer than in the elliptic case.

Our main results concerning the p-adic theory are the following. We define a p-adic modular form as an equivalence class of Cauchy sequences of classical modular forms of level μ_N, with N prime to p, with respect to an appropriate p-adic topology; see 10.8. A choice of cusp allows one to identify a p-adic modular form with a p-adic uniform limit of q-expansions of classical modular forms of a fixed level μ_N; see 10.10 and 10.11. Thus, this definition is in the spirit of Serre's original definition in the one dimensional case [Se] and we are able to present a theory very similar to that of loc. cit. In particular, a p-adic modular form has a well defined weight in a p-adic completion $\widehat{\mathbb{X}}$ of the group of universal characters $\mathbb{X}$; see 10.11.

On the other hand, one may define p-adic modular forms as regular functions on a formal scheme along lines established by Katz [Ka2], [Ka4] (who works mostly in the unramified case); see 11.4. The group $(O_L \otimes_{\mathbf{Z}} \mathbf{Z}_p)^*$ acts on this ring of regular functions and we interpret $\widehat{\mathbb{X}}$ as p-adic characters of $(O_L \otimes_{\mathbf{Z}} \mathbf{Z}_p)^*$. We prove that the notions of a modular form in Serre's approach and Katz' approach agree under minor restrictions: one may then identify a modular form of weight $\chi \in \widehat{\mathbb{X}}$ with a regular function transforming under the action of the group $(O_L \otimes_{\mathbf{Z}} \mathbf{Z}_p)^*$ by the character χ; see 11.11.

Using this, we extend the definition of the theta operators defined modulo p to p-adic operators that agree with those in Katz [Ka4] when defined; see 12.15, 12.23 and 12.26. This allows us to present many examples of p-adic modular forms; see 12.27.

The arithmetic of (p-adic) Hilbert modular forms has already seen some major achievements through Hida's theory and the association of Galois representations to Hilbert modular forms by A. Wiles and R. Taylor. The subject continues to develop rapidly and as the final version of this manuscript is written, related work by P. Kassaei on p-adic modular forms over Shimura curves, by A. Abbes-A. Mokrane, M. Kisin-K. F. Lai and E. Nevens on canonical subgroups of abelian varieties, and by F. Diamond and collaborators on the Serre conjectures for Hilbert modular forms are presumed to appear soon. The connection and mutual applications between these theories and the theory exposed below is yet to be explored and many interesting problems still need to be resolved. Among which, the construction of analytic families and eigenvarieties of Hilbert modular forms, the issue of boundedness from below of filtrations of Hilbert modular forms, a theory of theta cycles and the applications of the canonical subgroup to over-convergence. The authors hope to return to some of these topics in a future work.

2 Notations.

2.1 *Notation.* Let L be a totally real number field. Let p be a prime of $\mathbf{Z}$. Let

- O_L be the ring of integers of L;
- $g = [L : \mathbf{Q}]$ be the degree of L over $\mathbf{Q}$;
- D_L be the different ideal of L over $\mathbf{Q}$ and d_L be the discriminant;
- $\{(\mathfrak{I}, \mathfrak{I}^+)\}$ be fractional O_L-ideals, with the natural notion of positivity, forming a set of representatives of the strict class group of L i. e., of the isomorphism classes of projective O_L-modules of rank 1 with a notion of positivity;
- K be a Galois closure of L. Fix embeddings $K \hookrightarrow \overline{\mathbf{Q}}_p$ and $K \hookrightarrow \mathbf{C}$;
- $\{\gamma\colon L \to K\}$ be the set of distinct embeddings of L into K;
- $\{\mathfrak{P} | \, \mathfrak{P} \text{ divides } p\}$ be the primes of O_L above p. For each prime $\mathfrak{P}$ over p, let $\pi_{\mathfrak{P}} \in O_L$ be a generator of $\mathfrak{P} \otimes_{\mathbf{Z}} \mathbf{Z}_p$ such that $\pi_{\mathfrak{P}} \notin \mathfrak{P}'$ for any other prime $\mathfrak{P}'$ over p different from $\mathfrak{P}$;
- $L_{\mathfrak{P}}$ be the completion of L at $\mathfrak{P}$ and $O_{L_{\mathfrak{P}}}$ be its ring of integers;
- $e_{\mathfrak{P}}$ be the ramification index of $\mathfrak{P}$ over p;
- $\mathrm{k}_{\mathfrak{P}} = O_L/\mathfrak{P}$ and $f_{\mathfrak{P}} = [\mathrm{k}_{\mathfrak{P}} : \mathbf{F}_p]$. Denote by $\mathbf{W}(\mathrm{k}_{\mathfrak{P}})$ the Witt vectors of $\mathrm{k}_{\mathfrak{P}}$.

Let k be a perfect field of characteristic p such that k contains isomorphic copies of the residue fields $\{\mathrm{k}_{\mathfrak{P}}\}_{\mathfrak{P}|p}$ of the prime ideals of O_L over p. Let $\sigma\colon k \to k$ be the *absolute* Frobenius on k sending $x \mapsto x^p$. Let $\sigma\colon \mathbf{W}(k) \to \mathbf{W}(k)$ be the Frobenius automorphism – the unique lift of σ to the Witt vectors $\mathbf{W}(k)$ of k. For each prime $\mathfrak{P}$ of O_L over p let

$$\{\bar{\sigma}_{\mathfrak{P},i}\colon \mathrm{k}_{\mathfrak{P}} = O_L/\mathfrak{P} \longrightarrow k\}_{i=1,\ldots,f_{\mathfrak{P}}}$$

$$(\text{resp. } \{\hat{\sigma}_{\mathfrak{P},i}\colon \mathbf{W}(\mathrm{k}_{\mathfrak{P}}) \longrightarrow \mathbf{W}(k)\}_{i=1,\ldots,f_{\mathfrak{P}}})$$

be the set of different homomorphisms from the residue field $\mathrm{k}_{\mathfrak{P}}$ of O_L at $\mathfrak{P}$ to k (resp. from $\mathbf{W}(\mathrm{k}_{\mathfrak{P}})$ to $\mathbf{W}(k)$) ordered so that

$$\sigma \circ \bar{\sigma}_{\mathfrak{P},i} = \bar{\sigma}_{\mathfrak{P},i+1} \qquad (\text{resp.} \quad \sigma \circ \hat{\sigma}_{\mathfrak{P},i} = \hat{\sigma}_{\mathfrak{P},i+1}).$$

Then

$$O_L \underset{\mathbf{Z}}{\otimes} \mathbf{W}(k) = \prod_{\mathfrak{P}} O_{L_{\mathfrak{P}}} \underset{\mathbf{Z}_p}{\otimes} \mathbf{W}(k) = \prod_{\mathfrak{P}} \Big(\prod_{i=1}^{f_{\mathfrak{P}}} O_{L_{\mathfrak{P}}} \underset{\mathbf{W}(\mathrm{k}_{\mathfrak{P}})}{\otimes} \mathbf{W}(k) \Big),$$

where for each prime $\mathfrak{P}$ the last isomorphism is induced by applying $O_{L_{\mathfrak{P}}} \otimes_{\mathbf{W}(\mathrm{k}_{\mathfrak{P}})} (\cdot)$ to the isomorphism

$$(\hat{\sigma}_{\mathfrak{P},1}, \ldots, \hat{\sigma}_{\mathfrak{P},f_{\mathfrak{P}}}) \colon \mathbf{W}(\mathrm{k}_{\mathfrak{P}}) \underset{\mathbf{Z}_p}{\otimes} \mathbf{W}(k) \xrightarrow{\sim} \prod_{i=1}^{f_{\mathfrak{P}}} \mathbf{W}(k).$$

For each prime $\mathfrak{P}$ and for $1 \le i \le f_{\mathfrak{P}}$, denote by

$$\mathfrak{e}_{\mathfrak{P},i} \in O_L \underset{\mathbf{Z}}{\otimes} \mathbf{W}(k)$$

the associated idempotent.

Let $\mathfrak{p}$ be a prime of O_K over p and fix an embedding $O_K/\mathfrak{p} \hookrightarrow k$. Let σ be a lifting to $O_{K_{\mathfrak{p}}}$ of the *absolute* Frobenius on $O_K/\mathfrak{p}$. For every prime $\mathfrak{P}$ of O_L over p let

$$\{\sigma_{\mathfrak{P},i} \colon O_{L_{\mathfrak{P}}} \longrightarrow O_{K_{\mathfrak{p}}}\}_{i=1,\ldots,f_{\mathfrak{P}}}$$

be extensions of the homomorphisms $\{\hat{\sigma}_{\mathfrak{P},i}\}_{i=1,\ldots,f_{\mathfrak{P}}}$ to homomorphisms from the $\mathfrak{P}$-adic completion $O_{L_{\mathfrak{P}}}$ of O_L to the $\mathfrak{p}$-adic completion $O_{K_{\mathfrak{p}}}$ of O_K.

2.2 Definition. *Let*

$$\mathcal{G} := \mathrm{Res}_{O_L/\mathbf{Z}}(\mathbf{G}_{m,O_L}) \colon \mathrm{Schemes} \longrightarrow \mathrm{Groups}$$

be the Weil restriction of $\mathbf{G}_{m,O_L}$ *to* $\mathbf{Z}$ *i. e., the functor associating to a scheme* S *the group* $\big(\Gamma(S, O_S) \otimes_{\mathbf{Z}} O_L\big)^*$. *If* T *is a scheme, we write*

$$\mathcal{G}_T := \mathcal{G} \underset{\mathrm{Spec}(\mathbf{Z})}{\times} T.$$

If $T = \mathrm{Spec}(R)$, *we write* $\mathcal{G}_R$ *for* $\mathcal{G}_T$. *For any scheme* T *define*

$$\mathbb{X}_T := \mathrm{Hom}_{\mathrm{GR}}(\mathcal{G}_T, \mathbf{G}_{m,T})$$

as the group of characters of $\mathcal{G}_T$. *We often write* $\mathbb{X}$ *for* $\mathbb{X}_{O_K}$.

3 Moduli spaces of abelian varieties with real multiplication.

3.1 Note carefully. Throughout this paper we fix a fractional ideal $\mathfrak{I}$ with its natural positive cone $\mathfrak{I}^+$, among the ones chosen in 2.1. Below, we discuss Hilbert moduli spaces and Hilbert modular forms, where the polarization datum is fixed and equal to $(\mathfrak{I}, \mathfrak{I}^+)$. Our notation, though, does not reflect that. When we are compelled to consider the same notions with all the polarization modules chosen in 2.1 simultaneously, this will be explicitly mentioned.

3.2 Definition. *Let* S *be a scheme. Let* N *be a positive integer. Denote by*

$$\mathfrak{M}(S, \mu_N) \to S$$

the moduli stack over S of $\mathfrak{J}$-polarized abelian varieties with real multiplication by O_L and μ_N-level structure. It is a fibred category over the category of S-schemes. If T is a scheme over S, the objects of the stack over T are the $\mathfrak{J}$-polarized Hilbert-Blumenthal abelian schemes over T relative to O_L with μ_N-level structure i. e., quadruples $(A, \iota, \lambda, \varepsilon)$ consisting of

a) *an abelian scheme $A \to T$ of relative dimension g;*

b) *an O_L-action i. e., a ring homomorphism*

$$\iota\colon O_L \hookrightarrow \mathrm{End}_T(A);$$

c) *a polarization*

$$\lambda\colon (M_A, M_A^+) \xrightarrow{\sim} (\mathfrak{J}, \mathfrak{J}^+)$$

i. e., an O_L-linear isomorphism of sheaves on the étale site of T between the invertible O_L-module M_A of symmetric O_L-linear homomorphisms from A to its dual $A^\vee$ and the ideal $\mathfrak{J}$ of L, identifying the positive cone of polarizations M_A^+ with $\mathfrak{J}^+$;

d) *an O_L-linear injective homomorphism*

$$\varepsilon\colon \mu_N \underset{\mathbf{Z}}{\otimes} \mathrm{D}_L^{-1} \hookrightarrow A,$$

where for any scheme S over T we define

$$(\mu_N \underset{\mathbf{Z}}{\otimes} \mathrm{D}_L^{-1})(S) := \mu_N(S) \underset{\mathbf{Z}}{\otimes} \mathrm{D}_L^{-1}.$$

We require that the following condition, called the Deligne-Pappas condition, holds:

(DP) *the morphism $A \otimes_{O_L} M_A \longrightarrow A^t$ is an isomorphism.*

Let p be a prime. Suppose that S is a scheme over $\mathbf{Z}_p$. Denote by

$$\mathfrak{M}(S, \mu_N)^{\mathrm{ord}}$$

the open substack of $\mathfrak{M}(S, \mu_N)$ whose objects are the $\mathfrak{J}$-polarized abelian schemes, with real multiplication by O_L and with μ_N-level structure, that are geometrically ordinary.

3.3 Remark. If $N \geq 4$, the level structure of type μ_N is rigid [Go2, Lem. 1.1] and hence $\mathfrak{M}(S, \mu_N)$ is represented by a scheme over S.

Furthermore, $\mathfrak{M}(S, \mu_N)$ is geometrically irreducible over S. To see this one may reduce to the case S is the spectrum of $\mathbf{Z}$ localized at a prime p. By the theory of local models cf. [DePa], the geometric fibers are locally irreducible. Thus, it suffices to prove that the fibers are geometrically connected.

Consider first the case N prime to p. Replace the scheme $\mathfrak{M}(\mathbf{Z}_{(p)}, \mu_N)$ with its Satake compactification, X. It is normal, projective and flat over $\mathbf{Z}_{(p)}$ [Ch]. One then argues that in the Stein factorization $X \to Y \to \mathrm{Spec}(\mathbf{Z}_{(p)})$, the scheme Y is integral. By Zariski's main theorem and complex uniformization it follows that $Y = \mathrm{Spec}(\mathbf{Z}_{(p)})$ and, hence, that the geometric fibers of $X \to \mathrm{Spec}(\mathbf{Z}_{(p)})$ are connected. Since $\mathfrak{M}(\mathbf{F}_p, \mu_N)$ is dense in $X \otimes \mathbf{F}_p$, it is also geometrically connected.

It remains to consider the covering $\mathfrak{M}(\mathbf{F}_p, \mu_{p^r N}) \to \mathfrak{M}(\mathbf{F}_p, \mu_N)^{\mathrm{ord}}$. The irreducibility of $\mathfrak{M}(\mathbf{F}_p, \mu_{p^r N})$ follows from a monodromy argument as in [Ri, §III].

3.4 $\Gamma_0(\mathfrak{D})$*-level structure.* Let N be an integer and let $\mathfrak{D}$ be an ideal of O_L prime to N. Let T be a scheme. A $\mathfrak{I}$-polarized abelian scheme over T with real multiplication by O_L and μ_N and $\Gamma_0(\mathfrak{D})$-level structures is a quintuple

$$(A, \iota, \lambda, \varepsilon, H)/T,$$

where $(A, \iota, \lambda, \varepsilon)$ is a $\mathfrak{I}$-polarized abelian scheme over T with real multiplication by O_L and μ_N-level structure, and H is an O_L-invariant closed subgroup scheme

$$H \hookrightarrow A,$$

such that H is isomorphic to the constant group scheme $(O_L/\mathfrak{D})$ locally étale on T.

If S is a scheme, let

$$\mathfrak{M}(S, \mu_N, \Gamma_0(\mathfrak{D})) \longrightarrow S$$

be the moduli stack over S of $\mathfrak{I}$-polarized abelian varieties with real multiplication by O_L and $\mu_N \times \Gamma_0(\mathfrak{D})$-level structure. If T is an S-scheme, the objects of the stack over T consist of $\mathfrak{I}$-polarized abelian schemes over T with real multiplication by O_L and μ_N and $\Gamma_0(\mathfrak{D})$-level structures.

3.5 Definition. *Let T be a scheme. We say that an abelian scheme $\pi\colon A \to T$ with O_L-action $\iota\colon O_L \hookrightarrow \mathrm{End}_T(A)$ satisfies the Rapoport condition if*

(R) $\pi_*\Omega^1{}_{A/T}$ *is a locally free $O_T \otimes_{\mathbf{Z}} O_L$-module.*

Let S be a scheme. Denote by

$$\mathfrak{M}(S, \mu_N)^{\mathrm{R}}$$

the open substack of $\mathfrak{M}(S, \mu_N)$ whose objects are the $\mathfrak{I}$-polarized abelian schemes with real multiplication by O_L, with μ_N-level structure and satisfying the Rapoport condition.

3.6 Remark. We make several observations concerning the Rapoport condition.

1) A quadruple $(A, \iota, \lambda, \varepsilon)$ as in (a)-(d) of 3.2 satisfying (R) *automatically* satisfies (DP). It follows from [Ra, Cor. 1.13] and the characteristic 0 theory.
2) If d_L is invertible in S, then (R) and (DP) are equivalent. See [DePa, Cor. 2.9].
3) Suppose that S is a scheme over $\mathbf{Z}_p$. A quadruple $(A, \iota, \lambda, \varepsilon)$ as in (a)-(d) of 3.2, such that $A \to S$ is geometrically ordinary, *automatically* satisfies (R). Using that (R) is an open condition, one reduces to the case $S = \mathrm{Spec}(k)$ where k is an algebraically closed field of characteristic p. Now, $\mathrm{H}^1_{\mathrm{crys}}(A/\mathbf{W}(k))$ is a free module of rank 2 over $O_L \otimes \mathbf{W}(k)$ [Ra, Lem. 1.3] that can be identified with the Dieudonné module of the p-divisible group $A[p^\infty]$. Since A is ordinary, $A[p^\infty]$ decomposes as $A[p^\infty]^0 \oplus A[p^\infty]^{\mathrm{et}}$ and so $\mathrm{H}^1_{\mathrm{crys}}(A/\mathbf{W}(k))$ decomposes as the sum of the respective Dieudonné modules M^0 and M^{et}. Since A admits an O_L-polarization by [Ra, Prop. 1.12], the $O_L \otimes \mathbf{W}(k)$-modules M^0, M^{et} are isogenous, hence isomorphic (this uses that $O_L \otimes \mathbf{W}(k)$ is a product of discrete valuation rings). It follows that M^0 is a free $O_L \otimes \mathbf{W}(k)$-module. Since $M^0/pM^0 \cong \mathrm{H}^0(A, \Omega^1_{A/k})$ as $O_L \otimes k$-modules, the Rapoport condition holds.
4) We have open immersions

$$\mathfrak{M}(S, \mu_N)^{\mathrm{ord}} \hookrightarrow \mathfrak{M}(S, \mu_N)^{\mathrm{R}} \hookrightarrow \mathfrak{M}(S, \mu_N).$$

Fiberwise over S, the complement of $\mathfrak{M}(S,\mu_N)^{\mathrm{R}}$ in $\mathfrak{M}(S,\mu_N)$ has codimension at least 2. See [DePa, Prop. 4.4].

3.7 Remark. Let N_1 and N_2 be positive integers such that $N_2|N_1$ and $N_2 \geq 4$. The morphism

$$\mathfrak{M}(S,\mu_{N_1}) \longrightarrow \mathfrak{M}(S,\mu_{N_2})$$

is étale. Its image is open. Let $s \in S$ be a geometric point with residue field of characteristic l. Consider the induced étale morphism

$$\mathfrak{M}(k(s),\mu_{N_1}) \longrightarrow \mathfrak{M}(k(s),\mu_{N_2}).$$

Then

- if $l|N_1$, but $l \not| N_2$, the map is not surjective and its image coincides with the ordinary locus;
- otherwise the map is surjective.

Assume that R is a local artinian ring with residue field of characteristic l. Suppose that $N_1 N_2^{-1} = l^k$. Then, the morphism from $\mathfrak{M}(S,\mu_{N_1})$ to its image in $\mathfrak{M}(S,\mu_{N_2})$ is Galois with Galois group isomorphic to

$$\mathrm{Aut}_{O_L}\big(\mu_{N_1 N_2^{-1}} \underset{\mathbf{Z}}{\otimes} \mathrm{D}_L^{-1}\big) \simeq \big(O_L/N_1 N_2^{-1} O_L\big)^* = \big(O_L/l^k O_L\big)^*.$$

4 Properties of $\mathcal{G}$.

The notation is as in 2.2. The characters of $\mathcal{G}$ over a scheme S are the *weights* of Hilbert modular forms over S. For this reason we study them in some detail in this section. Special emphasis is reserved to positive characteristic. An important phenomenon is that if p ramifies in L, there exist characters over O_K (a normal extension of O_L) which coincide in characteristic p. This explains why the kernel of the q-expansion map in characteristic p is generated by g relations if p is inert in L and by a single relation if p is totally ramified. We also investigate the existence of "exotic" weights (as opposed to the universal ones) over artinian local rings such as Witt vectors of finite length.

4.1 Definition. *(The universal characters) The homomorphism of O_K-algebras*

$$O_L \underset{\mathbf{Z}}{\otimes} O_K \xrightarrow{\prod_\gamma \gamma \otimes \mathrm{id}} \prod_{\gamma: L \to K} O_K$$

defines a morphism of O_K-schemes

$$\mathcal{G}_{O_K} \xrightarrow{\prod_\gamma \chi_\gamma} \prod_{\gamma: L \to K} \mathbf{G}_{m,O_K}.$$

The characters $\{\chi_\gamma | \gamma: L \to K\}$ are called the basic characters of $\mathcal{G}_{O_K}$. Denote by

$$\mathbb{X}^\dagger_{O_K} \subset \mathbb{X}_{O_K} := \mathrm{Hom}_{O_K}\big(\mathcal{G}_{O_K}, \mathbf{G}_{m,O_K}\big)$$

the cone spanned by the basic characters. We also use the notation

$$\mathbb{X}^\dagger = \mathbb{X}^\dagger_{O_K}, \qquad \mathbb{X} = \mathbb{X}_{O_K}.$$

Define **Nm** *to be the character which is the product of the basic ones.*

Let T be a scheme over O_K. The subgroup $\mathbb{X}_T^U$ of the character group $\mathbb{X}_T$ of $\mathcal{G}_T$, spanned by the basic characters, is called the subgroup of universal characters.

4.2 *The generic fiber.* The group scheme $\mathcal{G}_{\mathbf{Q}}$ is a *torus*. Let K be the fixed Galois closure of L over $\mathbf{Q}$, then the universal characters induce an isomorphism

$$\mathcal{G}_K \xrightarrow{\prod_\gamma \chi_\gamma} \prod_{\gamma: L\to K} \mathbf{G}_{m,K}.$$

In fact, $O_L\otimes_{\mathbf{Z}} K \xrightarrow{\prod_\gamma \gamma\otimes \mathrm{id}} \prod_{\gamma:L\to K} K$ is an isomorphism.

4.3 *The group scheme $\mathcal{G}_{\mathbf{Z}_p}$.* We have

$$\mathcal{G}_{\mathbf{Z}_p} \xrightarrow{\sim} \prod_{\mathfrak{P}|p} \mathcal{G}^{\mathfrak{P}},$$

where $\mathcal{G}^{\mathfrak{P}} = \mathrm{Res}_{O_{L_\mathfrak{P}}/\mathbf{Z}_p}\big(\mathbf{G}_{m,O_{L_\mathfrak{P}}}\big)$ for all primes $\mathfrak{P}$ of O_L over p. This follows from the isomorphism $O_L\otimes_{\mathbf{Z}}\mathbf{Z}_p \xrightarrow{\sim} \prod_{\mathfrak{P}|p} O_{L_\mathfrak{P}}$. For each $\mathfrak{P}$ the natural inclusion $\mathbf{W}(\mathrm{k}_\mathfrak{P}) \to O_{L_\mathfrak{P}}$ defines a closed immersion of group schemes

$$\mathrm{Res}_{\mathbf{W}(\mathrm{k}_\mathfrak{P})/\mathbf{Z}_p}\big(\mathbf{G}_{m,\mathbf{W}(\mathrm{k}_\mathfrak{P})}\big) \hookrightarrow \mathcal{G}^{\mathfrak{P}}.$$

The group scheme $\mathrm{Res}_{\mathbf{W}(\mathrm{k}_\mathfrak{P})/\mathbf{Z}_p}\big(\mathbf{G}_{m,\mathbf{W}(\mathrm{k}_\mathfrak{P})}\big)$ is a torus. Let k be a field containing the fields $\mathrm{k}_\mathfrak{P}$ for all primes $\mathfrak{P}$. Then for every $\mathfrak{P}$ we have canonical isomorphisms

$$\mathrm{Res}_{\mathbf{W}(\mathrm{k}_\mathfrak{P})/\mathbf{Z}_p}\big(\mathbf{G}_{m,\mathbf{W}(\mathrm{k}_\mathfrak{P})}\big) \times_{\mathrm{Spec}(\mathbf{Z}_p)} \mathrm{Spec}\big(\mathbf{W}(k)\big) \xrightarrow{\sim} \prod_{\mathrm{k}_\mathfrak{P}\to k} \mathbf{G}_{m,\mathbf{W}(k)} \xrightarrow{\sim} \mathbf{G}_{m,\mathbf{W}(k)}^{f_\mathfrak{P}}.$$

4.4 *The special fiber of $\mathcal{G}_{\mathbf{Z}_p}$.* Let $\mathfrak{P}$ be a prime of O_L over p. We have an exact sequence

$$0 \longrightarrow \mathcal{G}_{\mathbf{F}_p}^{\mathfrak{P},u} \longrightarrow \mathcal{G}_{\mathbf{F}_p}^{\mathfrak{P}} \longrightarrow \mathcal{G}_{\mathbf{F}_p}^{\mathfrak{P},ss} \longrightarrow 0,$$

where $\mathcal{G}_{\mathbf{F}_p}^{\mathfrak{P},u}$ stands for the maximal unipotent subgroup of $\mathcal{G}_{\mathbf{F}_p}^{\mathfrak{P}}$ and $\mathcal{G}_{\mathbf{F}_p}^{\mathfrak{P},ss}$ is the semisimple part. The natural inclusion $\mathrm{k}_\mathfrak{P} \hookrightarrow O_{L_\mathfrak{P}}/pO_{L_\mathfrak{P}}$ defines a subgroup scheme

$$\mathrm{Res}_{\mathrm{k}_\mathfrak{P}/\mathbf{F}_p}\big(\mathbf{G}_{m,\mathrm{k}_\mathfrak{P}}\big) \hookrightarrow \mathcal{G}_{\mathbf{F}_p}^{\mathfrak{P}}$$

mapping isomorphically to $\mathcal{G}_{\mathbf{F}_p}^{\mathfrak{P},ss}$. In particular, the exact sequence above is split.

If k is a field containing $\mathrm{k}_\mathfrak{P} = O_L/\mathfrak{P}$, then

$$\mathrm{Res}_{\mathrm{k}_\mathfrak{P}/\mathbf{F}_p}\big(\mathbf{G}_{m,\mathrm{k}_\mathfrak{P}}\big) \times_{\mathrm{Spec}(\mathbf{F}_p)} \mathrm{Spec}(k) \xrightarrow{\sim} \prod_{\mathrm{k}_\mathfrak{P}\to k} \mathbf{G}_{m,k} \xrightarrow{\sim} \mathbf{G}_{m,k}^{f_\mathfrak{P}}$$

is a split torus. In particular, if k contains all the residue fields of O_K at all primes above p, the reduction map

$$\mathbb{X}_{O_K} \longrightarrow \mathbb{X}_k$$

is surjective and, hence, $\mathbb{X}_k$ is spanned by the universal characters. As shown in Example 4.7, the reduction map is not an isomorphism in general.

4.5 Remark. We have

$$\mathbb{X}_k = \prod_{\mathfrak{P}} \left(\prod_{i=1,\ldots,f_{\mathfrak{P}}} \chi_{\mathfrak{P},i}^{\mathbf{Z}} \right),$$

where $\chi_{\mathfrak{P},i}$ is the character of the torus $\mathcal{G}_{\overline{\mathbf{F}}_p}^{\mathfrak{P},ss} \cong \mathrm{Res}_{\mathrm{k}_{\mathfrak{P}}/\mathbf{F}_p}(\mathbf{G}_{m,\mathrm{k}_{\mathfrak{P}}})$ defined over k by the homomorphism $\bar{\sigma}_{\mathfrak{P},i}$ of 2.1.

4.6 *Example: the inert case.* Suppose that p is inert in L. Then the morphisms defined in 4.1 are isomorphisms after tensoring with $\mathbf{Z}_p$. In particular, $\mathcal{G}_{\mathbf{Z}_p}$ is a torus. The natural morphisms of character groups

$$\mathbb{X}_{\overline{\mathbf{F}}_p} \longleftarrow \mathbb{X}_{O_K} \longrightarrow \mathbb{X}_K$$

are all isomorphisms.

4.7 *Example: the totally ramified case.* Suppose that the prime p is totally ramified in L and $g > 1$. Let $\mathfrak{P}$ be the unique prime above p and fix an isomorphism $O_L/pO_L \cong \mathrm{k}_{\mathfrak{P}}[T]/(T^g)$. Then

$$\left(\mathcal{G}_{\overline{\mathbf{F}}_p}^{\mathfrak{P}}\right)(R) = \left(R \otimes \mathrm{k}_{\mathfrak{P}}[T]/(T^g)\right)^*$$

for any $\mathrm{k}_{\mathfrak{P}}$-algebra R. In particular, the toric part of $\mathcal{G}_{\overline{\mathbf{F}}_p}^{\mathfrak{P}}$ is one dimensional. The natural morphism of character groups

$$\mathbb{X}_{O_K} \longrightarrow \mathbb{X}_{\overline{\mathbf{F}}_p}$$

has a non-trivial kernel. In particular, all the basic characters have the same reduction Ψ.

4.8 *Exotic characters of* $\mathcal{G}$. Let k be a perfect field of positive characteristic p. Let $\mathbf{W}_{m+1}(k) \to \mathbf{W}_m(k)$ be the canonical surjective homomorphism on truncated Witt vectors. It is defined by a principal ideal (ϵ) satisfying $p\epsilon = 0$. Consider the induced reduction homomorphism

$$\alpha\colon \mathbb{X}_{\mathbf{W}_{m+1}(k)} \longrightarrow \mathbb{X}_{\mathbf{W}_m(k)}.$$

Let

$$\chi\colon \mathcal{G}_{\mathbf{W}_{m+1}(k)} \longrightarrow \mathbf{G}_{m,\mathbf{W}_{m+1}(k)}$$

be a character. Write

$$\mathcal{G}_{\mathbf{W}_{m+1}(k)} := \mathrm{Spec}(A), \qquad \mathbf{G}_{m,\mathbf{W}_{m+1}(k)} = \mathrm{Spec}\left(\mathbf{W}_{m+1}(k)\left[t, \frac{1}{t}\right]\right).$$

Then

$$\chi \in \mathrm{Ker}(\alpha) \qquad \Longleftrightarrow \qquad \chi(t) = 1 + \epsilon a, \quad a \in A/pA$$

with a satisfying

$$\Delta(a) = a \otimes 1 + 1 \otimes a, \qquad \mathfrak{u}(a) = 0,$$

where Δ (resp. $\mathfrak{u}$) is the comultiplication (resp. counit) of A. Hence the kernel of α is

$$\mathrm{Ker}(\alpha) = \mathrm{Hom}_{k-\mathrm{GR}}(\mathcal{G}_k, \mathbf{G}_{a,k}).$$

Let $\mathcal{G}_k^u$ be the unipotent part of $\mathcal{G}_k$. As soon as it is non-trivial, the group $\mathrm{Ker}(\alpha)$ is not finitely generated. In fact, there exists a surjective homomorphism $\mathcal{G}_k \to \mathbf{G}_{a,k}$ and it is acted upon by

$$\mathrm{End}_{k-\mathrm{GR}}\big(\mathbf{G}_{a,k}\big) \xrightarrow{\sim} \big\{f \in k[X] \,|\, f \in k[X^p],\ f(0) = 0\big\},$$

where the group structure on the right hand side is induced by composition. In particular, as soon as p is ramified, $\mathbb{X}_{\mathbf{W}_{m+1}(k)}$ is not finitely generated and hence is not generated by the basic characters.

The reader may have noticed that this phenomenon cannot occur for $g = 1$. The phenomenon of exotic weights, and the exotic Hilbert modular forms associated to them, may only occur for $g > 1$ in the presence of ramification. However, this "pathology" is a peculiarity of artinian bases and disappears in the truly p-adic situation. The situation is different when we consider the characters that lift to $\mathbf{W}_m(k)$ for all integers m.

4.9 Definition. *(The formal characters) Let $\widehat{\mathcal{G}}$ be the smooth formal group over $\mathrm{Spf}(\mathbf{Z}_p)$ associated to the group scheme $\mathcal{G}$. Let $\widehat{O}_K$ be the completion of O_K with respect to the ideal pO_K. Define*

$$\mathbb{X}_{\widehat{O}_K}\left(\widehat{\mathcal{G}}\right) = \mathrm{Hom}_{\widehat{O}_K}\left(\widehat{\mathcal{G}}\widehat{\times}_{\mathrm{Spf}\left(\mathbf{Z}_p\right)}\mathrm{Spf}\big(\widehat{O}_K\big), \widehat{\mathbf{G}}_{m,\widehat{O}_K}\right)$$

as the group of formal characters of $\widehat{\mathcal{G}}$ over $\mathrm{Spf}(\widehat{O}_K)$. For any character $\chi \in \mathbb{X}_{O_K}$ define $\widehat{\chi}$ to be the induced element of $\mathbb{X}_{\widehat{O}_K}(\widehat{\mathcal{G}})$.

4.10 Proposition. *The natural morphism*

$$\prod_{\gamma: L \to K} \widehat{\chi}_\gamma^{\mathbf{Z}} \longrightarrow \mathbb{X}_{\widehat{O}_K}\left(\widehat{\mathcal{G}}\right)$$

is an isomorphism.

Proof: Consider the following natural diagram over $\widehat{O}_K$

$$\begin{array}{ccc}
\mathcal{G}_{\widehat{O}_K} & \longleftarrow & \mathcal{G}_{\widehat{K}} \\
\uparrow & & \uparrow \\
\mathcal{G}_{\widehat{O}_K}^{\backslash p} & & \mathcal{G}_{\widehat{K}}^{\backslash \mathbf{1}} \\
\uparrow & & \downarrow \\
\mathcal{G}_{\widehat{O}_K}^{\backslash p, \mathbf{1}} & \xrightarrow{\xi} & \mathcal{G}_{\widehat{O}_K}^{\backslash \mathbf{1}},
\end{array}$$

where $\mathcal{G}_{\widehat{O}_K}^{\backslash \mathbf{1}}$ (resp. $\mathcal{G}_{\widehat{K}}^{\backslash \mathbf{1}}$) stands for the completion of $\mathcal{G}_{\widehat{O}_K}$ (resp. $\mathcal{G}_{\widehat{K}}$) along the identity section, where $\mathcal{G}_{\widehat{O}_K}^{\backslash p}$ stands for the completion of $\mathcal{G}_{\widehat{O}_K}$ along the special fiber $\mathcal{G}_{\widehat{O}_K/p\widehat{O}_K}$ and where $\mathcal{G}_{\widehat{O}_K}^{\backslash p,\mathbf{1}}$ stands for the completion of $\mathcal{G}_{\widehat{O}_K}$ at the identity section of $\mathcal{G}_{\widehat{O}_K/p\widehat{O}_K}$. Let $\mathcal{G} = \mathrm{Spec}(A)$. Then

a) using that A is a domain and Krull's theorem, one deduces that all the arrows in the diagram are inclusions at the level of the algebras of regular functions;

b) since $\mathcal{G}$ is smooth over $\mathrm{Spec}(\mathbf{Z})$, we have that $\mathcal{G}^{\backslash \mathbf{1}}_{\widehat{O}_K}$ is the formal spectrum of a power series ring over $\widehat{O}_K$. Hence, the associated algebra is p-adically complete. In particular, the natural map $\xi\colon \mathcal{G}^{\backslash p,\mathbf{1}}_{\widehat{O}_K} \to \mathcal{G}^{\backslash \mathbf{1}}_{\widehat{O}_K}$ induces an isomorphism at the level of the algebras of regular functions;

c) under this isomorphism the underlying diagram of algebras is commutative.

A similar diagram exists replacing $\mathcal{G}$ with $\mathbf{G}_{m,\mathbf{Z}}$. To any element of $\mathbb{X}_{\widehat{O}_K}(\widehat{\mathcal{G}})$, one can associate a formal character of $\mathcal{G}^{\backslash p}_{\widehat{O}_K}$ and hence, by the diagram above and b), an element in

$$\mathbb{X}_{\widehat{K}}\left(\mathcal{G}^{\backslash \mathbf{1}}_{\widehat{K}}\right) := \mathrm{Hom}_{\widehat{K}}\left(\mathcal{G}^{\backslash \mathbf{1}}_{\widehat{K}}, \mathbf{G}^{\backslash \mathbf{1}}_{m,\widehat{K}}\right).$$

On the other hand to any character of $\mathcal{G}_K$ (resp. of $\mathcal{G}$), one can associate an element in $\mathbb{X}_{\widehat{K}}(\mathcal{G}^{\backslash \mathbf{1}}_{\widehat{K}})$ (resp. in $\mathbb{X}_{\widehat{O}_K}(\widehat{\mathcal{G}})$). In particular, we obtain a commutative diagram

$$\begin{array}{ccc} \mathbb{X} & \xrightarrow{\sim} & \mathbb{X}_K \\ \downarrow & & \downarrow t \\ \mathbb{X}_{\widehat{O}_K}(\widehat{\mathcal{G}}) & \xrightarrow{s} & \mathbb{X}_{\widehat{K}}\left(\mathcal{G}^{\backslash \mathbf{1}}_{\widehat{K}}\right). \end{array}$$

By 4.2, we have $\mathcal{G}_K \cong \prod_{\gamma: K \to L} \mathbf{G}_{m,K}$. Hence, t is an isomorphism. By (a) the map s is injective. Hence, all the arrows in the diagram of characters are isomorphisms. This proves the claim.

Let B be a ring with an ideal $\mathfrak{p}$ such that $B/\mathfrak{p}$ is of characteristic p. In the rest of this section we define a certain filtration $\{\mathbb{X}_{B,\mathfrak{p}}(n)\}$ on the group of characters $\mathbb{X}_B$ of $\mathcal{G} \times_{\mathrm{Spec}(\mathbf{Z})} \mathrm{Spec}(B)$. If B is a p-adic ring and an O_K-algebra, the topology of $\mathbb{X}_B$ induced by this filtration is separated and we study the resulting completion $\widehat{\mathbb{X}}_{B,\mathfrak{p}}$. The motivation is that p-adic modular forms "à la Serre" over B, defined in 10.8, have a well defined weight in $\widehat{\mathbb{X}}_{B,\mathfrak{p}}$.

4.11 Definition. *Let B be a ring with an ideal $\mathfrak{p}$ such that $B/\mathfrak{p}$ is of characteristic p. For any non-negative integer n define*

$$\mathbb{X}_{B,\mathfrak{p}}(n) := \left\{\chi \in \mathbb{X}_B \,|\, \chi\colon (O_L/p^n O_L)^* \longrightarrow (B/\mathfrak{p}^n B)^* \text{ is trivial}\right\}.$$

Define

$$\widehat{\mathbb{X}}_{B,\mathfrak{p}} := \lim_{\infty \leftarrow n} \mathbb{X}_B/\mathbb{X}_{B,\mathfrak{p}}(n).$$

We suppress the index $\mathfrak{p}$ if no confusion is likely to arise.

4.12 *The structure of $\widehat{\mathbb{X}}_{B,\mathfrak{p}}$.* Suppose that B is an O_K-algebra and that $\mathfrak{p}$ is a maximal ideal. Note that $\mathbb{X}_B/\mathbb{X}_{B,\mathfrak{p}}(1)$ is isomorphic to $\mathbb{X}_{B/\mathfrak{p}B}/\mathbb{X}_{B/\mathfrak{p}B}(1)$. It follows from 4.4 that $\mathbb{X}_B/\mathbb{X}_{B,\mathfrak{p}}(1)$ consists of universal characters and

$$\mathbb{X}_B/\mathbb{X}_{B,\mathfrak{p}}(1) \xrightarrow{\sim} \mathrm{Hom}\big((O_L/pO_L)^*, (B/\mathfrak{p})^*\big) = \prod_{\mathfrak{P}|p} \mathrm{Hom}\big(\mathrm{k}_{\mathfrak{P}}^*, (B/\mathfrak{p})^*\big).$$

Since

$$\mathrm{Ker}\Big(\big(B/\mathfrak{p}^{n+1}\big)^* \to \big(B/\mathfrak{p}^n\big)^*\Big) = 1 + \mathfrak{p}^n \pmod{\mathfrak{p}^{n+1}} \cong \mathfrak{p}^n/\mathfrak{p}^{n+1},$$

it is killed by p. In particular, for any $n \geq 1$ we have that p kills $\mathbb{X}_{B,\mathfrak{p}}(n)/\mathbb{X}_{B,\mathfrak{p}}(n+1)$.

Suppose that B is flat over O_K. We aim to prove that

$$\cap_n \mathbb{X}_{B,\mathfrak{p}}(n) = \{1\}.$$

Let $\widehat{B} := \lim_n B/\mathfrak{p}^n$. Let $\sigma_1, \ldots, \sigma_g \colon L \to K$ denote the distinct embeddings of L in K. Let $\chi_1, \ldots, \chi_g$ be the associated basic characters of $\mathrm{Res}_{O_K/\mathbf{Z}}(\mathbf{G}_{m,O_K})$ as in 4.1. The flatness of B over $\mathbf{Z}$ implies that $\mathbb{X}_B = \chi_1^{\mathbf{Z}} \times \cdots \times \chi_g^{\mathbf{Z}}$; see 4.13. Suppose $\chi \in \cap_n \mathbb{X}_{B,\mathfrak{p}}(n)$. Write $\chi = \chi_1^{a_1} \cdots \chi_g^{a_g}$. By assumption the homomorphism $\chi \colon (O_L \otimes_{\mathbf{Z}} \mathbf{Z}_p)^* \to \widehat{B}^*$ is trivial. Let $U \subset (O_L \otimes_{\mathbf{Z}} \mathbf{Z}_p)^*$ and $V \subset O_L \otimes_{\mathbf{Z}} \mathbf{Z}_p$ be subgroups of finite index where the exponential map is defined and induces an isomorphism $\exp \colon V \xrightarrow{\sim} U$. The map $\log_{\widehat{B}} \circ \chi \circ \exp \colon V \longrightarrow \widehat{B}$, given by $l \otimes z \longmapsto \sum_{i=1}^{g} a_i \sigma_i(l) z$, is the zero map and factors through the completion of K at the prime ideal $\mathfrak{p} \cap O_K$. The independence of the embeddings $\sigma_1, \ldots, \sigma_g$ gives that $a_i = 0$ for every $i = 1, \ldots, g$ i. e., that $\chi = 1$ as claimed.

4.13 Lemma. *Suppose that B is a ring flat over O_K. Then*

1. *we have*
$$\mathbb{X}_B = \mathbb{X}_{O_K} = \prod_{\gamma : L \to K} \chi_\gamma^{\mathbf{Z}};$$
2. *the topology on $\mathbb{X}_B$ induced by the system of subgroups $\{\mathbb{X}_{B,\mathfrak{p}}(n)\}_{n \in \mathbf{N}}$ is separated, i. e.*
$$\cap_{n \in \mathbf{N}} \mathbb{X}_{B,\mathfrak{p}}(n) = \{1\};$$
3. *finally, $\widehat{\mathbb{X}}_{B,\mathfrak{p}}$ is independent of $\mathfrak{p}$ and is the product of*
   - *a finite group of order prime to p isomorphic to $\prod_{\mathfrak{P}|p} \mathrm{Hom}(\mathrm{k}_{\mathfrak{P}}^*, (B/\mathfrak{p})^*)$;*
   - *a topological group isomorphic to $\mathbf{Z}_p^g$.*

Proof: The flatness implies that $\mathbb{X}_B \hookrightarrow \mathbb{X}_{B \otimes_{\mathbf{Z}} \mathbf{Q}}$. Claim (1) follows from 4.1. Claim (2) follows from 4.12. Note that $\mathbb{X}_{B,\mathfrak{p}}(1)$ is a free abelian group of rank g. Consider

$$\widehat{\mathbb{X}}_{B,\mathfrak{p}}(1) := \lim_{\infty \leftarrow n} \mathbb{X}_{B,\mathfrak{p}}(1)/\mathbb{X}_{B,\mathfrak{p}}(n).$$

By 4.12 the group $\mathbb{X}_{B,\mathfrak{p}}(n)/\mathbb{X}_{B,\mathfrak{p}}(n+1)$ is killed by p for $n \geq 1$. Hence $p^n \mathbb{X}_{B,\mathfrak{p}}(1)$ is contained in $\mathbb{X}_{B,\mathfrak{p}}(n)$. We obtain a continuous surjective homomorphism

$$\lim_{\infty \leftarrow n} \mathbb{X}_{B,\mathfrak{p}}(1)/p^n \mathbb{X}_{B,\mathfrak{p}}(1) \longrightarrow \widehat{\mathbb{X}}_{B,\mathfrak{p}}(1).$$

The group on the left hand side is isomorphic to $\mathbf{Z}_p^g$ as topological groups. It follows from (2) that $\widehat{\mathbb{X}}_{B,\mathfrak{p}}(1)$ has no p-torsion. Hence, $\widehat{\mathbb{X}}_{B,\mathfrak{p}}(1)$ is isomorphic to $\mathbf{Z}_p^g$. We conclude from 4.12 that $\widehat{\mathbb{X}}_{B,\mathfrak{p}} = \widehat{\mathbb{X}}_{B,\mathfrak{p}}(1) \times \mathbb{X}_{B,\mathfrak{p}}/\mathbb{X}_{B,\mathfrak{p}}(1)$. This proves claim (3).

4.14 *Examples.* In the inert case $\widehat{\mathbb{X}}_{B,\mathfrak{p}}$ is the direct product of a cyclic subgroup of order $p^{f_{\mathfrak{p}}} - 1$ and a free $\mathbf{Z}_p$-module with basis $\{\chi_1^p \chi_2^{-1}, \ldots, \chi_g^p \chi_1^{-1}\}$. In the totally ramified case $\widehat{\mathbb{X}}_{B,\mathfrak{p}}$ is the direct product of a cyclic subgroup of order $p-1$ and a free $\mathbf{Z}_p$-module with basis $\left\{\chi_1^{p-1}, \chi_2 \chi_1^{-1}, \ldots, \chi_g \chi_1^{-1}\right\}$ (where one can obtain a similar basis with any χ_i in place of χ_1).

5 Hilbert modular forms.

5.1 Definition. *Let S be a scheme. Let χ be an element of $\mathbb{X}_S$. A $\mathfrak{J}$-polarized Hilbert modular form f over S of weight χ and level μ_N is a rule associating to*

i. *any affine scheme* $\mathrm{Spec}(R)$ *over* S;
ii. *any $\mathfrak{J}$-polarized Hilbert-Blumenthal variety $(A, \iota, \lambda, \varepsilon)$ over* $\mathrm{Spec}(R)$ *with μ_N-level structure;*
iii. *any generator ω of $\Omega^1_{A/R}$ as $R \otimes_{\mathbf{Z}} O_L$-module,*

an element $f(A, \iota, \lambda, \varepsilon, \omega)$ of R i. e.,

$$(A, \iota, \lambda, \varepsilon, \omega) \longmapsto f(A, \iota, \lambda, \varepsilon, \omega),$$

with the following properties:

I. *the value $f(A, \iota, \lambda, \varepsilon, \omega)$ depends only on the isomorphism class over* $\mathrm{Spec}(R)$ *of the $\mathfrak{J}$-polarized Hilbert-Blumenthal variety $(A, \iota, \lambda, \varepsilon, \omega)$ with μ_N-level structure and with section ω;*

II. *the rule f is compatible with base change i. e., if $\phi\colon R \to B$ is a ring homomorphism*

$$f\left((A, \iota, \lambda, \varepsilon, \omega) \underset{\mathrm{Spec}(R)}{\times} \mathrm{Spec}(B)\right) = \phi\big(f(A, \iota, \lambda, \varepsilon, \omega)\big);$$

III. *if $\alpha \in \mathcal{G}(R) = (R \otimes_{\mathbf{Z}} O_L)^*$, then*

$$f(A, \iota, \lambda, \varepsilon, \alpha^{-1}\omega) = \chi(\alpha) f(A, \iota, \lambda, \varepsilon, \omega).$$

Denote by $\mathbf{M}(S, \mu_N, \chi)$ the $\Gamma(S, O_S)$-module of such functions.

5.2 Remark. The notation is as in 3.4. One defines $\mathfrak{J}$-polarized Hilbert modular forms of weight χ and level $\mu_N \times \Gamma_0(\mathfrak{D})$ in the obvious way. Denote by $\mathbf{M}(S, \mu_N, \Gamma_0(\mathfrak{D}), \chi)$ the $\Gamma(S, O_S)$-module of such functions.

5.3 Remark. The formation of the space of modular forms does not commute with base change. In fact, it is not even true that, given a morphism of schemes $T \to S$, the map

$$\mathbf{M}(S, \mu_N, \chi) \underset{\Gamma(S, O_S)}{\otimes} \Gamma(T, O_T) \longrightarrow \mathbf{M}(T, \mu_N, \chi)$$

is surjective. For example, one can take $S = \mathrm{Spec}(O_L)$ and $T = \mathrm{Spec}(\mathrm{k}_{\mathfrak{P}})$ for some prime ideal $\mathfrak{P}$ of O_L. By [vdG, Lem. I.6.3], the weight $\chi = \prod_\gamma \chi_\gamma^{a_\gamma}$ of non-zero modular forms over O_L must satisfy the condition that $a_\gamma \geq 0$. See 4.1 for the notation. Examples show that this is not necessary if T is of characteristic p. See 7.12.

5.4 Definition. *Suppose that $N \geq 4$. Let R be a ring and let $\chi \in \mathbb{X}_R$. By 3.3 there exists a universal $\mathfrak{J}$-polarized Hilbert-Blumenthal abelian scheme with μ_N-level structure*

$$(\mathrm{A}^{\mathrm{U}}, \iota^{\mathrm{U}}, \lambda^{\mathrm{U}}, \varepsilon^{\mathrm{U}})$$

over $\mathfrak{M}(\mathbf{Z},\mu_N)$. The pull-back of the sheaf $\pi_*\Omega^1_{\mathrm{A^U}/\mathfrak{M}(\mathbf{Z},\mu_N)^{\mathrm{R}}}$ to the Rapoport locus $\mathfrak{M}(R,\mu_N)^{\mathrm{R}}$, defined in 3.5, is a locally free $O_{\mathfrak{M}(R,\mu_N)^{\mathrm{R}}}\otimes_{\mathbf{Z}} O_L$-module of rank 1. Hence, it defines a cohomology class

$$c\in \mathrm{H}^1\left(\mathfrak{M}(R,\mu_N)^{\mathrm{R}},\left(O_{\mathfrak{M}(R,\mu_N)^{\mathrm{R}}}\underset{\mathbf{Z}}{\otimes} O_L\right)^*\right).$$

Let $\mathcal{L}_\chi$ be the invertible sheaf over $\mathfrak{M}(R,\mu_N)^{\mathrm{R}}$ associated to the cohomology class defined by the push-forward of c via the map χ

$$\mathrm{H}^1\left(\mathfrak{M}(R,\mu_N)^{\mathrm{R}},\left(O_{\mathfrak{M}(R,\mu_N)^{\mathrm{R}}}\underset{\mathbf{Z}}{\otimes} O_L\right)^*\right)\stackrel{\chi}{\longrightarrow}\mathrm{H}^1\left(\mathfrak{M}(R,\mu_N)^{\mathrm{R}},O^*_{\mathfrak{M}(R,\mu_N)^{\mathrm{R}}}\right).$$

See [Ra, §6.8].

5.5 Proposition. 1. *To define a $\mathfrak{I}$-polarized Hilbert modular form f of weight χ and level μ_N over a ring R is equivalent to define a section of $\mathcal{L}_\chi$ over $\mathfrak{M}(R,\mu_N)^{\mathrm{R}}$.*
2. *For $\chi,\chi'\in\mathbb{X}_R$ we have a canonical isomorphism*

$$\mathcal{L}_\chi\otimes\mathcal{L}_{\chi'}\stackrel{\sim}{\longrightarrow}\mathcal{L}_{\chi\chi'}$$

as invertible sheaves on $\mathfrak{M}(R,\mu_N)^{\mathrm{R}}$.

5.6 Remark. By definition, Hilbert modular forms are only defined over the Rapoport locus. It is interesting to note that although one can make sense of Hilbert modular forms of parallel weight (i. e., $\mathbf{Nm}^k$) over the whole moduli space, this is not possible for non parallel weight. For example, when p is maximally ramified in O_L the invertible sheaf $\mathcal{L}_\psi$ (ψ any basic character) does not extend as an invertible sheaf over the whole moduli space [AG].

6 The q-expansion map.

6.1 *Uniformization of semiabelian schemes with O_L-action.* Let R be a noetherian normal excellent domain complete and separated with respect to an ideal I such that $\sqrt{I}=I$. Let $S:=\mathrm{Spec}(R)$ and $S_0=\mathrm{Spec}(R/I)$. Let $S\backslash S_0$ be the open subscheme defined by the complement of S_0 in S. Then we have an equivalence of categories between

(O_L-DD) the category of 1-motives

$$\underline{q}\colon \mathfrak{B}\longrightarrow \mathfrak{A}^{-1}\mathrm{D}_L^{-1}\underset{\mathbf{Z}}{\otimes}\mathbf{G}_m(S\backslash S_0),$$

where

- $\mathfrak{A}$ and $\mathfrak{B}$ are projective O_L-modules of rank 1;
- $\underline{q}$ is O_L-linear;
- the O_L-module $M:=\mathfrak{A}\mathfrak{B}$ is endowed with a notion of positivity so that $(\mathfrak{A}\mathfrak{B}^{-1},(M\mathfrak{B}^{-2})^+)\cong(\mathfrak{I},\mathfrak{I}^+)$.

The motive is subject to the *degeneration condition* that for any $m=ab\in\mathfrak{A}\mathfrak{B}$ such that $m\in M^+$ the element of the fraction field of A associated to $a(\underline{q}(b))\in\mathbf{G}_m(S\backslash S_0)$ belongs to I. Here, we identify $\mathfrak{A}$ with the character group of the torus $\mathfrak{A}^{-1}\mathrm{D}_L^{-1}\otimes_{\mathbf{Z}}\mathbf{G}_{m,\mathbf{Z}}$.

The morphisms from an object $\underline{q}\colon \mathfrak{B} \longrightarrow \mathfrak{A}^{-1}\mathrm{D}_L^{-1} \otimes_{\mathbf{Z}} \mathbf{G}_m\big(S\backslash S_0\big)$ to a second object $\underline{q}'\colon \mathfrak{B}' \longrightarrow (\mathfrak{A}')^{-1}\mathrm{D}_L^{-1} \otimes_{\mathbf{Z}} \mathbf{G}_m\big(S\backslash S_0\big)$ are defined by the O_L-linear maps

$$\big(\mathfrak{A}\big)^{-1} \longrightarrow \big(\mathfrak{A}'\big)^{-1}$$

such that in

$$\begin{array}{ccc} \mathfrak{B} & \xrightarrow{\underline{q}} & \mathfrak{A}^{-1}\mathrm{D}_L^{-1} \otimes_{\mathbf{Z}} \mathbf{G}_m\big(S\backslash S_0\big) \\ & & \big\downarrow \\ \mathfrak{B}' & \xrightarrow{\underline{q}'} & (\mathfrak{A}')^{-1}\mathrm{D}_L^{-1} \otimes_{\mathbf{Z}} \mathbf{G}_m\big(S\backslash S_0\big) \end{array}$$

there exists a left vertical arrow making the diagram commute. Note that the degeneration condition implies that $\underline{q}$ and $\underline{q}'$ are injective. Hence, if such map exists it is unique.

(O_L-Deg) the category whose objects consist of semiabelian schemes G over S endowed with an O_L-action such that

- $G \times_S \big(S\backslash S_0\big)$ is an abelian scheme with O_L-action and with polarization module isomorphic to $(\mathfrak{I}, \mathfrak{I}^+)$;
- $G \times_S S_0 \cong \mathfrak{A}\mathrm{D}_L^{-1} \otimes_{\mathbf{Z}} \mathbf{G}_{m,S_0}$, where $\mathfrak{A}$ is a projective O_L-module of rank 1.

The morphisms are the homomorphisms as semiabelian schemes commuting with the O_L-action.

The semiabelian scheme G associated to a 1-motive $\underline{q}\colon \mathfrak{B} \to \mathfrak{A}^{-1}\mathrm{D}_L^{-1} \otimes_{\mathbf{Z}} \mathbf{G}_m(S\backslash S_0)$ is usually denoted by

$$\left(\mathfrak{A}^{-1}\mathrm{D}_L^{-1} \underset{\mathbf{Z}}{\otimes} \mathbf{G}_{m,S}\right)/\underline{q}(\mathfrak{B}).$$

See [Ra] for details.

6.2 Remark. We gather some properties of this construction.

i) We have an isomorphism :

$$\left(\left(\mathfrak{A}^{-1}\mathrm{D}_L^{-1} \underset{\mathbf{Z}}{\otimes} \mathbf{G}_{m,S}\right)/\underline{q}(\mathfrak{B})\right) \underset{S}{\times} S_0 \cong \mathfrak{A}^{-1}\mathrm{D}_L^{-1} \underset{\mathbf{Z}}{\otimes} \mathbf{G}_{m,S_0}.$$

ii) For any integral O_L-ideal $\mathfrak{D}$ we have an exact sequence of group schemes over $S\backslash S_0$

$$0 \to \left(\frac{\mathfrak{A}}{\mathfrak{D}\mathfrak{A}}\right)(1) \longrightarrow \left(\mathfrak{A}^{-1}\mathrm{D}_L^{-1} \underset{\mathbf{Z}}{\otimes} \mathbf{G}_{m,S}/\underline{q}(\mathfrak{B})\right)\big[\mathfrak{D}\big] \longrightarrow \mathfrak{D}^{-1}\mathfrak{B}/\mathfrak{B} \to 0,$$

where $(\mathfrak{A}/\mathfrak{D}\mathfrak{A})(1)$ is the Cartier dual of the constant group scheme $\mathfrak{A}/\mathfrak{D}\mathfrak{A}$.

iii) The module of invariant differentials $\omega_{G/S}$ of G relative to S satisfies

$$\omega_{G/S} \xrightarrow{\sim} \left((\mathfrak{A}^{-1}\mathrm{D}_L^{-1})^{\vee} \underset{\mathbf{Z}}{\otimes} R\right)\frac{dt}{t} \xrightarrow{\sim} \left(\mathfrak{A} \underset{\mathbf{Z}}{\otimes} R\right)\frac{dt}{t},$$

where $R\,dt/t$ is the module of invariant differentials of $\mathbf{G}_{m,S}$ relative to S.

iv) The O_L-module of symmetric O_L-linear homomorphisms from the abelian scheme $G \times_S \big(S\backslash S_0\big)$ to its dual is canonically isomorphic to

$$\mathrm{Hom}_{O_L}\big(\mathfrak{B}, (\mathfrak{A}^{-1}\mathrm{D}_L^{-1})^{\vee}\big) = \mathfrak{B}^{-1}\mathfrak{A} = M\mathfrak{B}^{-2} = M^{-1}\mathfrak{A}^2$$

and it is endowed with a notion of positivity induced by the one on M.

6.3 *Tate objects.* Fix projective O_L-modules $\mathfrak{A}$ and $\mathfrak{B}$ of rank 1. Fix a notion of positivity on $M := \mathfrak{A}\mathfrak{B}$ such that $\big(\mathfrak{A}\mathfrak{B}^{-1}, (M\mathfrak{B}^{-2})^+\big) \cong (\mathfrak{I}, \mathfrak{I}^+)$. Fix a rational polyhedral cone decomposition $\{\sigma_\beta\}_\beta$ of the dual cone to $M^+_{\mathbf{R}} \subset M_{\mathbf{R}}$, which is invariant under the action of the totally positive units of O_L and such that, modulo this action, the number of polyhedra is finite. Let

$$S := M^\vee \underset{\mathbf{Z}}{\otimes} \mathbf{G}_{m,\mathbf{Z}}.$$

For any σ_β we obtain an affine torus embedding $S \subset S_{\sigma_\beta}$. Let $S^\wedge_{\sigma_\beta}$ be the spectrum of the ring obtained by completing the affine scheme S_{σ_β} along the closed subscheme $S_{\sigma_\beta,0} = S_{\sigma_\beta}\backslash S$ with reduced structure. Over the base S_{σ_β} one has a canonical 1-motive giving rise over $S^\wedge_{\sigma_\beta}\backslash S_{\sigma_\beta,0}$ to a $\mathfrak{I}$*-polarized generalized Tate object*

$$\mathbf{Tate}(\mathfrak{A},\mathfrak{B})_{\sigma_\beta} = \Big(\mathfrak{A}^{-1}\mathrm{D}_L^{-1} \underset{\mathbf{Z}}{\otimes} \mathbf{G}_{m,S^\wedge_{\sigma_\beta}}/\underline{q}(\mathfrak{B})\Big) \underset{S^\wedge_{\sigma_\beta}}{\times} \big(S^\wedge_{\sigma_\beta}\backslash S_{\sigma_\beta,0}\big).$$

See [Ra] or [Ka1, §1.1]. Define

$$\mathbf{Z}\big((\mathfrak{A},\mathfrak{B},\sigma_\beta)\big) := \mathbf{Z}((q^\nu))_{\nu\in\sigma_\beta}.$$

It can be interpreted as

$$\mathrm{Spec}\Big(\mathbf{Z}\big((\mathfrak{A},\mathfrak{B},\sigma_\beta)\big)\Big) = S^\wedge_{\sigma_\beta}\backslash S_{\sigma_\beta,0}.$$

6.4 *Unramified cusps.* A $\mathfrak{I}$-polarized unramified cusp of level μ_N over $\mathrm{Spec}(R)$ is a quadruple $(\mathfrak{A},\mathfrak{B},\varepsilon,\mathrm{j})$, where

a) $\mathfrak{A}$ and $\mathfrak{B}$ are fractional ideals such that $\mathfrak{A}\mathfrak{B}^{-1} = \mathfrak{I}$;
b) $\varepsilon\colon N^{-1}O_L/O_L \xrightarrow{\sim} N^{-1}\mathfrak{A}^{-1}/\mathfrak{A}^{-1}$ is an O_L-linear isomorphism;
c) $\mathrm{j}\colon \mathfrak{A}\otimes_{\mathbf{Z}} R \xrightarrow{\sim} O_L \otimes_{\mathbf{Z}} R$ is an $O_L\otimes_{\mathbf{Z}} R$-linear isomorphism.

6.5 Remark. i) By 6.2, the equality $\mathfrak{A}\mathfrak{B}^{-1} = \mathfrak{I}$ identifies $\mathfrak{I}$ with $M_{\mathbf{Tate}(\mathfrak{A},\mathfrak{B})_{\sigma_\beta}}$, the group of O_L-linear symmetric homomorphisms from $\mathbf{Tate}(\mathfrak{A},\mathfrak{B})_{\sigma_\beta}$ to its dual, and the cone $\mathfrak{I}^+$ with the cone $M^+_{\mathbf{Tate}(\mathfrak{A},\mathfrak{B})_{\sigma_\beta}}$ of polarizations. Thus, we obtain a canonical polarization datum

$$\lambda_{\mathrm{can}}\colon \Big(M_{\mathbf{Tate}(\mathfrak{A},\mathfrak{B})_{\sigma_\beta}}, M^+_{\mathbf{Tate}(\mathfrak{A},\mathfrak{B})_{\sigma_\beta}}\Big) \xrightarrow{\sim} (\mathfrak{I},\mathfrak{I}^+).$$

ii) By 6.2, the isomorphism ε defines a canonical μ_N-level structure on the abelian scheme $\mathbf{Tate}(\mathfrak{A},\mathfrak{B})_{\sigma_\beta}$.
iii) The isomorphism j defines a canonical isomorphism

$$\Big(\Omega^1_{\mathbf{Tate}(\mathfrak{A},\mathfrak{B})_{\sigma_\beta}/\left(S^\wedge_{\sigma_\beta}\backslash S_{\sigma_\beta,0}\right)}\Big) \underset{\mathbf{Z}}{\otimes} R \xrightarrow{\sim} O_L \underset{\mathbf{Z}}{\otimes} \Big(R \underset{\mathbf{Z}}{\otimes} O_{S^\wedge_{\sigma_\beta}\backslash S_{\sigma_\beta,0}}\Big).$$

In some situations one has a canonical choice of j. For example:
a) if R is a $\mathbf{Q}$-algebra we get

$$\mathrm{j}_{\mathrm{can}}\colon \mathfrak{A}\underset{\mathbf{Z}}{\otimes} R \xrightarrow{\sim} (\mathfrak{A}\underset{\mathbf{Z}}{\otimes}\mathbf{Q})\underset{\mathbf{Q}}{\otimes} R \xrightarrow{\sim} L\underset{\mathbf{Q}}{\otimes} R = (O_L\underset{\mathbf{Z}}{\otimes}\mathbf{Q})\underset{\mathbf{Q}}{\otimes} R;$$

b) if $\mathfrak{A}$ is an integral O_L-ideal of norm invertible in R (e.g. $\mathfrak{A} = O_L$), then the natural inclusion $\mathfrak{A} \hookrightarrow O_L$ induces a canonical isomorphism

$$\mathrm{j}_{\mathrm{can}}\colon \mathfrak{A} \underset{\mathbf{Z}}{\otimes} R \xrightarrow{\sim} O_L \underset{\mathbf{Z}}{\otimes} R.$$

c) if $p^n R = 0$ and p^n divides N, then ε induces by duality an isomorphism

$$\mathfrak{A}/N\mathfrak{A} \xrightarrow{\sim} O_L/NO_L$$

and, consequently, a canonical isomorphism

$$\mathrm{j}_\varepsilon\colon \mathfrak{A} \underset{\mathbf{Z}}{\otimes} R \xrightarrow{\sim} O_L \underset{\mathbf{Z}}{\otimes} R.$$

6.6 *The q-expansion.* Let $\mathrm{Spec}(R)$ be an affine scheme. Let f be an element of $\mathbf{M}\big(R, \mu_N, \chi\big)$ i. e., a $\mathfrak{J}$-polarized Hilbert modular form over $\mathrm{Spec}(R)$ of weight χ and level μ_N. Denote

$$f\big(\mathbf{Tate}(\mathfrak{A}, \mathfrak{B}), \varepsilon, \mathrm{j}\big) := f\Big(\mathbf{Tate}(\mathfrak{A}, \mathfrak{B})_{\sigma_\beta} \underset{\mathrm{Spec}(\mathbf{Z})}{\times} \mathrm{Spec}(R), \lambda_{\mathrm{can}}, \varepsilon, \frac{dt}{t}\Big)$$

and call it the q-expansion of f at the unramified cusp $(\mathfrak{A}, \mathfrak{B}, \varepsilon, \mathrm{j})$.

6.7 *A variant.* Let $\mathfrak{D}$ be an integral O_L-ideal prime to N. A $\mathfrak{J}$-polarized cusp of level $\mu_N \times \Gamma_0(\mathfrak{D})$ over an affine scheme $\mathrm{Spec}(R)$ is a quintuple $(\mathfrak{A}, \mathfrak{B}, \varepsilon, H, \mathrm{j})$, where $(\mathfrak{A}, \mathfrak{B}, \varepsilon, \mathrm{j})$ satisfy (a)-(c) in 6.4 and

$$H \hookrightarrow \mathbf{Tate}(\mathfrak{A}, \mathfrak{B})_{\sigma_\beta} \underset{\mathrm{Spec}(\mathbf{Z})}{\times} \mathrm{Spec}(R)$$

denotes an O_L-invariant closed subgroup scheme isomorphic étale locally to the O_L-constant group scheme $\big(O_L/\mathfrak{D}\big)$.

Let f be an element of $\mathbf{M}\big(R, \mu_N, \Gamma_0(\mathfrak{D}), \chi\big)$ i. e., a $\mathfrak{J}$-polarized Hilbert modular form over $\mathrm{Spec}(R)$ of weight χ and level $\mu_N \times \Gamma_0(\mathfrak{D})$ in the sense of 5.2. Denote

$$f\big(\mathbf{Tate}(\mathfrak{A}, \mathfrak{B}), \varepsilon, H, \mathrm{j}\big) := f\Big(\mathbf{Tate}(\mathfrak{A}, \mathfrak{B})_{\sigma_\beta} \underset{\mathrm{Spec}(\mathbf{Z})}{\times} \mathrm{Spec}(R), \lambda_{\mathrm{can}}, \varepsilon, H, \frac{dt}{t}\Big)$$

and call it the q-expansion of f at the cusp $(\mathfrak{A}, \mathfrak{B}, \varepsilon, H, \mathrm{j})$.

6.8 Theorem. *Let $(\mathfrak{A}, \mathfrak{B}, \varepsilon, \mathrm{j})$ be an unramified cusp over R. Then the element $f\big(\mathbf{Tate}(\mathfrak{A}, \mathfrak{B}), \varepsilon, \mathrm{j}\big)$ does not depend on the cone decomposition $\{\sigma_\beta\}$. Moreover, it is of the form*

$$f\big(\mathbf{Tate}(\mathfrak{A}, \mathfrak{B}), \varepsilon, \mathrm{j}\big) = \sum_{\nu \in \mathfrak{A}\mathfrak{B}^+ \cup \{0\}} c_\nu q^\nu \in R[\![q^\nu]\!]_{\nu \in \mathfrak{A}\mathfrak{B}^+ \cup \{0\}}.$$

Proof: [Ra, §4.6] and [Ra, Prop. 4.9].

6.9 Remark. It is proven in [Ch, Thm. 4.3 (X)] that $R[\![q^\nu]\!]^{U^2}_{\nu \in \mathfrak{A}\mathfrak{B}^+ \cup \{0\}}$, the invariants under the action of squared units U^2 of O_L, is the completed local ring of the minimal compactification of $\mathfrak{M}\big(\mathrm{Spec}(R), \mu_N\big)$ at the cusp $(\mathfrak{A}, \mathfrak{B}, \varepsilon, \mathrm{j})$.

6.10 Theorem. *(q-expansion principle) Let R be a ring and let $(\mathfrak{A}, \mathfrak{B}, \epsilon, \mathrm{j})$ be an unramified cusp defined over R. Let f be an element of $\mathbf{M}(R, \mu_N, \chi)$.*

i. *If $f(\mathbf{Tate}(\mathfrak{A}, \mathfrak{B}), \varepsilon, \mathrm{j}) = 0$, then $f = 0$;*

ii. *If $(\mathfrak{A}, \mathfrak{B}, \epsilon, \mathrm{j})$ is defined over a subring R_0 of R and $f(\mathbf{Tate}(\mathfrak{A}, \mathfrak{B}), \varepsilon, \mathrm{j})$ belongs to $R_0[\![q^\nu]\!]_{\nu \in \mathfrak{A}\mathfrak{B}^+ \cup \{0\}}$, then f belongs to $\mathbf{M}(R_0, \mu_N, \chi)$.*

Proof: See [Ra, Thm. 6.7].

6.11 *The comparison with the complex theory.* Let

$$\sigma_1, \ldots, \sigma_g : L \longrightarrow \mathbf{R}$$

be the real embeddings of L. Let $\mathfrak{A}$ and $\mathfrak{B}$ be ideals of O_L such that $\mathfrak{J} = \mathfrak{A}\mathfrak{B}^{-1}$. Fix an O_L-linear isomorphism

$$\varepsilon : N^{-1}O_L/O_L \xrightarrow{\sim} N^{-1}\mathfrak{A}^{-1}/\mathfrak{A}^{-1}.$$

Define the group

$$\Gamma_N(\mathfrak{B} \oplus (\mathfrak{A}\mathrm{D}_L)^{-1}) := \Big\{ \begin{pmatrix} a & b \\ c & d \end{pmatrix} \mid a \in 1 + NO_L, d \in O_L, b \in (\mathfrak{B}\mathfrak{A}\mathrm{D}_L)^{-1},$$
$$c \in N\mathfrak{B}\mathfrak{A}\mathrm{D}_L, ad - bc = 1 \Big\}.$$

It acts on the g-fold product of the Poincaré upper half plane $\mathfrak{H}^g$ by

$$\begin{pmatrix} a & b \\ c & d \end{pmatrix} (z_1, \ldots, z_g) := \left(\ldots, \frac{\sigma_i(a) z_i + \sigma_i(b)}{\sigma_i(c) z_i + \sigma_i(d)}, \ldots \right)_{i=1,\ldots,g}.$$

The moduli space $\mathfrak{M}(\mathbf{C}, \mu_N)$ classifying $\mathfrak{J}$-polarized abelian varieties A over $\mathbf{C}$ with real multiplication by O_L and μ_N-level structure, in the sense of 3.2, is isomorphic, as an analytic manifold, to

$$\Gamma_N(\mathfrak{B} \oplus (\mathfrak{A}\mathrm{D}_L)^{-1}) \backslash \mathfrak{H}^g.$$

The abelian variety corresponding to $\tau \in \mathfrak{H}^g$ is

$$A_\tau := \mathbf{C}^g / (\mathfrak{B}\tau + (\mathfrak{A}\mathrm{D}_L)^{-1}).$$

The μ_N-level structure on A_τ is induced by ε. The vector space $\mathbf{M}(\mathbf{C}, \mu_N, \chi)$ of $\mathfrak{J}$-polarized modular forms of level μ_N and weight $\chi = \prod_{i=1}^g \chi_{\sigma_i}^{a_i}$ can be viewed, more classically, as the vector space of holomorphic functions

$$\begin{array}{ccc} \mathfrak{H}^g & \xrightarrow{f} & \mathbf{C} \\ \tau = (z_1, \ldots, z_g) & \mapsto & f(\tau), \end{array}$$

on which the action of the modular group $\Gamma_N(\mathfrak{B} \oplus (\mathfrak{A}\mathrm{D}_L)^{-1})$ is defined by the automorphic factor

$$j_\chi\Big(\mu, (z_1, \ldots, z_g)\Big) = \prod_{i=1}^g \big(\sigma_i(c) z_i + \sigma_i(d)\big)^{a_i} \qquad \text{with } \mu = \begin{pmatrix} a & b \\ c & d \end{pmatrix};$$

c.f. [vdG] or [Go3]. Fix a modular form f. Assuming $\mathfrak{B} = O_L$, one deduces that f, as a function, is invariant with respect to the translations on $\mathfrak{H}^g$

$$\tau \mapsto \tau + \alpha, \qquad \alpha \in (\mathfrak{A}\mathrm{D}_L)^{-1}.$$

In particular, it has q-expansion at the cusp $(i\infty,\dots,i\infty)$

$$f(\underline{q}) := a_0 + \sum_{\nu\in\mathfrak{A}^+} a_\nu q^\nu \qquad \text{with } q^\nu := \exp^{2\pi i\,\mathrm{Tr}_{L/\mathbf{Q}}(\nu\tau)},$$

where

$$\mathrm{Tr}_{L/\mathbf{Q}}(\nu\tau) := \sigma_1(\nu)z_1 + \ldots + \sigma_g(\nu)z_g.$$

By 6.2 we have a natural O_L-linear isomorphism

$$\mathrm{j}_{\mathrm{can}}\colon \mathfrak{A}\otimes_{\mathbf{Z}}\mathbf{C}\xrightarrow{\sim} O_L\otimes_{\mathbf{Z}}\mathbf{C}.$$

Note that under the exponentiation map $z\mapsto \exp^{2\pi i\,\mathrm{Tr}_{L/\mathbf{Q}}(z)}$, where we use the identification $\mathbf{C}^g = (\mathfrak{A}\mathrm{D}_L)^{-1}\otimes_{\mathbf{Z}}\mathbf{C}$, we have

$$\mathbf{C}^g/\big(\mathfrak{B}\tau + (\mathfrak{A}\mathrm{D}_L)^{-1}\big) \cong (\mathbf{C}^*)^g/\underline{q}(\mathfrak{B}),$$

with $\underline{q}(\mathfrak{B})$ equal to the image of $\mathfrak{B}\tau$. By the discussion in [Ra, §§6.13-6.15] it follows that

$$f(\underline{q}) = f\big(\mathbf{Tate}(\mathfrak{A},\mathfrak{B}),\varepsilon,\mathrm{j}_{\mathrm{can}}\big).$$

7 The partial Hasse invariants.

In this section we define a canonical set of Hilbert modular forms that we call "partial Hasse invariants". The name comes from the fact that these modular forms factor the Hasse invariant according to the O_L-structure. The partial Hasse invariants are defined in characteristic p and, in general, do not lift to characteristic zero. Indeed, because the weight of a non-cusp form in characteristic zero must be parallel, none of the partial Hasse invariants $h_{\mathfrak{P},i}$ defined below lifts to characteristic zero, unless p is maximally ramified in the totally real field L. In this case, there is a unique partial Hasse invariant $h_{\mathfrak{P}}$, which is a g-th root of the total Hasse invariant h (h is, up to a sign, the determinant of the Hasse-Witt matrix). In certain cases, see §20, one can guarantee the existence of a lift of a power of $h_{\mathfrak{P}}$; in general, the existence of such a lift, in particular to an Eisenstein series, is closely related to properties of special values of the p-adic zeta function of L and, hence, to Leopoldt's conjecture (cf. [Go1]).

The partial Hasse invariants, however, play a crucial role in the theory in several ways: they provide canonical generators for the kernel of the q-expansion map; they allow one to compactify the moduli scheme $\mathfrak{M}(k,\mu_{pN})$.

The results and the methods of this section appear already in [Go2] in the unramified case. In that case, the divisors of the partial Hasse invariants yield an interesting stratification of the moduli space $\mathfrak{M}(k,\mu_N)$, [Go1] and [GoOo].

7.1 *Notation.* Let k be a perfect field of characteristic p. Assume that it contains the residue fields $\mathrm{k}_{\mathfrak{P}} = O_L/\mathfrak{P}$ for all primes $\mathfrak{P}$ of O_L over p. Let R be a local ring with maximal ideal $\mathfrak{m}$ and residue field $R/\mathfrak{m} = k$.

For each representative $(\mathfrak{J},\mathfrak{J}^+)$ of the strict class group of L fixed in 2.1 choose an element of $\mathfrak{J}$ generating $\mathfrak{J}\otimes_{\mathbf{Z}}\mathbf{Z}_p$ as $O_L\otimes_{\mathbf{Z}}\mathbf{Z}_p$-module. It provides each $\mathfrak{J}$-polarized abelian scheme with real multiplication by O_L over a $\mathbf{Z}_p$-scheme with a polarization of degree prime to p.

7.2 Definition. *Let $n \geq m \geq 1$ be positive integers. Let N be an integer such that $N \geq 4$ and N is prime to p. Assume that R is an O_K-algebra. Define*

$$\mathfrak{M}(R/\mathfrak{m}^m, \mu_{p^n N}) \xrightarrow{\ \psi\ } \mathfrak{M}(R/\mathfrak{m}^m, \mu_N)^{\mathrm{ord}}$$

by

$$(A, \iota, \lambda, \varepsilon_{p^n N}) \longmapsto (A, \iota, \lambda, \varepsilon_N).$$

By 3.7, it is a Galois cover of smooth schemes over $R/\mathfrak{m}^m$ with group

$$\Gamma_n := \mathrm{Aut}_{R/\mathfrak{m}^m}\big(\mu_{p^n} \underset{\mathbf{Z}}{\otimes} \mathrm{D}_L^{-1}\big) = \big(O_L/p^n O_L\big)^*.$$

We can evaluate $\chi \in \mathbb{X}$ on Γ_n as follows: $\Gamma_n = \mathcal{G}(\mathbf{Z}/p^n\mathbf{Z}) \xrightarrow{\chi} (\mathbf{Z}/p^n\mathbf{Z})^* \to (R/\mathfrak{m}^m)^*$. Let

$$\mathfrak{M}(R/\mathfrak{m}^m, \mu_{p^n N})^{\mathrm{Kum}} \xrightarrow{\ \phi\ } \mathfrak{M}(R/\mathfrak{m}^m, \mu_N)^{\mathrm{ord}}$$

be the quotient of the above cover by the group of elements of $(O_L/p^n O_L)^*$ killed by

$$\big\{\chi\colon \big(O_L/p^n O_L\big)^* \to \big(R/\mathfrak{m}^m\big)^* \,|\, \chi \quad \text{is a universal character}\,\big\}.$$

Let $G_{m,n}$ be the Galois group of ϕ. By [DeRi, Thm. 4.5] *it acts transitively on the fibers of ϕ.*

7.3 Remark. If $m = n = 1$, then $G := G_{1,1} \xrightarrow{\sim} \prod_{\mathfrak{P}|p} \big(O_L/\mathfrak{P}\big)^*$.

7.4 Definition. *Let $\chi \in \mathbb{X}_{R/\mathfrak{m}^m}$ be a character in the sense of 2.2. Write*

$$\mu_{p^n} = \mathbf{Z}[t]/\big(t^{p^n} - 1\big).$$

Denote by

$$\frac{dt}{t} \in \Omega^1_{\mathrm{D}_L^{-1} \otimes_{\mathbf{Z}} \mu_{p^n}} \underset{\mathrm{Spec}(\mathbf{Z})}{\times} \mathrm{Spec}\big(\mathbf{Z}/p^n\mathbf{Z}\big)$$

the canonical generator, as a free $O_L/p^n O_L$-module of rank 1, of the submodule of invariant differentials of $\Omega^1_{\mathrm{D}_L^{-1} \otimes_{\mathbf{Z}} \mu_{p^n}} \times_{\mathrm{Spec}(\mathbf{Z})} \mathrm{Spec}\big(\mathbf{Z}/p^n\mathbf{Z}\big)$.

Suppose that $n \geq m$. Let

$$\pi\colon \big(\mathrm{A}^{\mathrm{U}}, \iota^{\mathrm{U}}, \lambda^{\mathrm{U}}, \varepsilon^{\mathrm{U}}_{p^n N}\big) \longrightarrow \mathfrak{M}(R/\mathfrak{m}^m, \mu_{p^n N})$$

be the $\mathfrak{J}$-polarized universal abelian scheme with real multiplication by O_L and $\mu_{p^n N}$-level structure. We use the notation:

$$\omega^{\mathrm{can}} := \varepsilon_{p^n N,*}\left(\frac{dt}{t}\right) \in \pi_* \Omega^1_{\mathrm{A}^{\mathrm{U}}/\mathfrak{M}\left(R/\mathfrak{m}^m, \mu_{p^n N}\right)}.$$

We can also view ω^{can} as a translation invariant differential in $\Omega^1_{\mathrm{A}^{\mathrm{U}}/\mathfrak{M}\left(R/\mathfrak{m}^m, \mu_{p^n N}\right)}$. See 7.5.

Define $a(\chi)$ as the unique $\mathfrak{J}$-polarized modular form of weight χ and level $\mu_{p^n N}$ over $R/\mathfrak{m}^m$ satisfying the following. Let T be a $R/\mathfrak{m}^m$-algebra. Let $(A, \iota, \lambda, \varepsilon)$ be a $\mathfrak{J}$-polarized Hilbert-Blumenthal abelian scheme over T with $\mu_{p^n N}$-level structure. Let ω be an $O_L \otimes_{\mathbf{Z}} T$-generator of $\mathrm{H}^0\big(A, \Omega^1_{A/T}\big)$. Then, there is a unique element γ of $\big(O_L \otimes_{\mathbf{Z}} T\big)^*$ such that

$$\omega = \gamma^{-1} \varepsilon_*\Big(\frac{dt}{t}\Big).$$

Define

$$a(\chi)\Big(A,\iota,\lambda,\varepsilon,\omega\Big) := \chi\ (\gamma)\,.$$

7.5 *Some explanations.* Decompose $\varepsilon^{\mathrm{U}}_{p^nN} := \varepsilon^{\mathrm{U}}_{p^n}\times\varepsilon^{\mathrm{U}}_{N}$. Since $n\geq m$, the embedding

$$\varepsilon^{\mathrm{U}}_{p^n}\colon \big(\mu_{p^n}\underset{\mathbf{Z}}{\otimes}\mathrm{D}_L^{-1}\big)\times\mathfrak{M}\big(R/\mathfrak{m}^m,\mu_{p^nN}\big)\to \mathrm{A}^{\mathrm{U}}$$

defines an isomorphism on tangent spaces at the origin relative to $\mathfrak{M}\big(R/\mathfrak{m}^m,\mu_{p^nN}\big)$. By duality, one gets a canonical isomorphism of the translation invariant differentials relative to $\mathfrak{M}\big(R/\mathfrak{m}^m,\mu_{p^nN}\big)$. This defines a canonical $O_L\otimes_{\mathbf{Z}} O_{\mathfrak{M}\left(R/\mathfrak{m}^m,\mu_{p^nN}\right)}$-generator of the translation invariant differentials of A^{U} over $\mathfrak{M}\big(R/\mathfrak{m}^m,\mu_{p^nN}\big)$:

$$\omega^{\mathrm{can}} := \varepsilon_{p^nN,*}\left(\frac{dt}{t}\right)\in\Omega^1_{\mathrm{A}^{\mathrm{U}}/\mathfrak{M}\left(R/\mathfrak{m}^m,\mu_{p^nN}\right)}.$$

By 5.4, the pull-back $\psi^*(\mathcal{L}_\chi)$ of the invertible sheaf $\mathcal{L}_\chi$, defined on $\mathfrak{M}\big(R/\mathfrak{m}^m,\mu_N\big)^{\mathrm{R}}$, to $\mathfrak{M}\big(R/\mathfrak{m}^m,\mu_{p^nN}\big)$ is obtained as the push-out of $\Omega^1_{\mathrm{A}^{\mathrm{U}}/\mathfrak{M}\left(R/\mathfrak{m}^m,\mu_{p^nN}\right)}$. The section ω^{can} of $\Omega^1_{\mathrm{A}^{\mathrm{U}}/\mathfrak{M}\left(R/\mathfrak{m}^m,\mu_{p^nN}\right)}$ defines by push-out a section of $\psi^*(\mathcal{L}_\chi)$. Using the equivalence between sections of $\psi^*\big(\mathcal{L}_\chi\big)$ and modular forms of weight χ and level μ_{p^nN} over $R/\mathfrak{m}^m$, we find that this section coincides with the modular form $a(\chi)$.

7.6 Proposition. *We have*

$$\Big(O_L\underset{\mathbf{Z}}{\otimes}O_{\mathfrak{M}\left(R/\mathfrak{m}^m,\mu_{p^nN}\right)}\Big)\,\omega^{\mathrm{can}} = \pi_*\Omega^1_{\mathrm{A}^{\mathrm{U}}/\mathfrak{M}\left(R/\mathfrak{m}^m,\mu_{p^nN}\right)}.$$

For any $\alpha\in\Gamma_n=\big(O_L/p^nO_L\big)^*$, *the induced action by pull-back*

$$\big[\alpha\big]\colon\pi_*\Omega^1_{\mathrm{A}^{\mathrm{U}}/\mathfrak{M}\left(R/\mathfrak{m}^m,\mu_{p^nN}\right)}\longrightarrow\pi_*\Omega^1_{\mathrm{A}^{\mathrm{U}}/\mathfrak{M}\left(R/\mathfrak{m}^m,\mu_{p^nN}\right)}$$

sends

$$\omega^{\mathrm{can}}\longmapsto\alpha^{-1}\,\omega^{\mathrm{can}}.$$

Proof: The first claim follows from 7.5. Decompose $\varepsilon^{\mathrm{U}}_{p^nN} := \varepsilon^{\mathrm{U}}_{p^n}\times\varepsilon^{\mathrm{U}}_{N}$. Since the universal $\mathfrak{I}$-polarized abelian scheme A^{U} over $\mathfrak{M}\big(R/\mathfrak{m}^m\mu_{p^nN}\big)$ is the pull-back of the universal $\mathfrak{I}$-polarized abelian scheme over $\mathfrak{M}\big(R/\mathfrak{m}^m,\mu_N\big)$, the automorphism α lifts to an automorphism of A^{U}, which we denote by α and induces the automorphism $[\alpha]$ on $\Omega^1_{\mathrm{A}^{\mathrm{U}}/\mathfrak{M}\left(R/\mathfrak{m}^m,\mu_{p^nN}\right)}$. By definition, $\alpha\big(\varepsilon^{\mathrm{U}}_{p^n}\big)=\big(\varepsilon^{\mathrm{U}}_{p^n}\circ 1\otimes\alpha\big)$. Hence,

$$\begin{aligned}\big(\varepsilon^{\mathrm{U}}_{p^n}\circ 1\otimes\alpha\big)_*\left(\frac{dt}{t}\right) &= \varepsilon^{\mathrm{U}}_{p^n,*}\left((1\otimes\alpha)\,\frac{dt}{t}\right)\\ &= \alpha^{-1}\omega^{\mathrm{can}}.\end{aligned}$$

7.7 Corollary. *For any* $\alpha\in\big(O_L/p^nO_L\big)^*$, *the induced action by pull-back,*

$$\big[\alpha\big]\colon\Gamma\Big(\mathfrak{M}\big(R/\mathfrak{m}^m,\mu_{p^nN}\big),\psi^*\big(\mathcal{L}_\chi\big)\Big)\longrightarrow\Gamma\Big(\mathfrak{M}\big(R/\mathfrak{m}^m,\mu_{p^nN}\big),\psi^*\big(\mathcal{L}_\chi\big)\Big),$$

maps

$$a(\chi) \longmapsto \chi^{-1}(\alpha)a(\chi).$$

7.8 Corollary. *The section ω^{can} descends to a section over $\mathfrak{M}(R/\mathfrak{m}^m, \mu_{p^n N})^{\text{Kum}}$ of the relative differentials of the pull-back by ϕ of the universal $\mathfrak{I}$-polarized abelian scheme over $\mathfrak{M}(R/\mathfrak{m}^m, \mu_N)$. We denote it ω^{can} by abuse of notation.*

The section $a(\chi)$ of $\psi^*(\mathcal{L}_\chi)$ descends to a section, denoted $a(\chi)$ by abuse of notation, of $\phi^*(\mathcal{L}_\chi)$ over $\mathfrak{M}(R/\mathfrak{m}^m, \mu_{p^n N})^{\text{Kum}}$. It is the push-out via χ of ω^{can}. If α belongs to the Galois group $G_{m,n}$ of ϕ, then

$$[\alpha](\omega^{\text{can}}) = \alpha^{-1}\omega^{\text{can}} \qquad \text{and} \qquad [\alpha]\big(a(\chi)\big) = \chi^{-1}(\alpha)a(\chi).$$

7.9 Proposition. 1. *The section $a(\chi)$ induces a trivialization of the invertible sheaf $\phi^*(\mathcal{L}_\chi)$ on $\mathfrak{M}(R/\mathfrak{m}^m, \mu_{p^n N})^{\text{Kum}}$.*

2. *For any universal characters χ and χ' the section $a(\chi) \otimes a(\chi')$ of $\mathcal{L}_\chi \otimes \mathcal{L}_{\chi'}$ is sent by the isomorphism defined in 5.5 to the section $a(\chi\chi')$ of $\mathcal{L}_{\chi\chi'}$.*

3. *For any universal character χ the q-expansion of $a(\chi)$ at any $\mathfrak{I}$-polarized cusp $(\mathfrak{A}, \mathfrak{B}, \varepsilon, \mathrm{j}_\varepsilon)$ (see 6.5 and 6.6) of level $\mu_{p^n N}$ over $R/\mathfrak{m}^m$ is 1.*

Proof: (1) and (2) are clear. Let $\mathbf{Tate}(\mathfrak{A}, \mathfrak{B})_{\sigma_\beta}$ be a Tate object with $\mu_{p^n N}$-level structure ε. The modular form $a(\chi)$ takes the value 1 on the section $\varepsilon_*(dt/t)$. The latter coincides with the section of the translation invariant relative differential defined by j_ε as in 6.5. This proves (3).

7.10 Remark. The Proposition justifies the introduction of the covering ϕ in 7.2.

7.11 *First construction of the Hasse invariants: the geometric definition.* Let R be a k-algebra. Let $(A, \iota, \lambda, \varepsilon)$ be a $\mathfrak{I}$-polarized Hilbert-Blumenthal abelian variety over R satisfying the condition (R) defined in 3.5. Let ω be an $O_L \otimes_{\mathbf{Z}} R$-basis of $\mathrm{H}^0(A, \Omega^1_{A/R})$. Using λ, and the choices in 7.1, we get an isomorphism

$$\mathrm{H}^0(A, \Omega^1_{A/R}) \xrightarrow{\sim} \mathrm{H}^0(A^\vee, \Omega^1_{A^\vee/R}).$$

Here $A^\vee$ is the dual abelian scheme. Via the canonical isomorphism

$$\mathrm{H}^0(A^\vee, \Omega^1_{A^\vee/R}) \xrightarrow{\sim} \mathrm{Hom}_R(\mathrm{H}^1(A, O_A), R),$$

the element ω determines a generator

$$\eta \in \mathrm{H}^1(A, O_A)$$

as $O_L \otimes_{\mathbf{Z}} R$-module.

The absolute Frobenius on A induces a σ-linear map $O_A \to O_A$ and consequently a σ-linear map

$$\mathrm{F}\colon \mathrm{H}^1(A, O_A) \longrightarrow \mathrm{H}^1(A, O_A).$$

Let $\mathfrak{P}$ be a prime of O_L above p and let $1 \le i \le f_{\mathfrak{P}}$. With the notation of 2.1, the σ-linearity of F implies that for each idempotent $\mathfrak{e}_{\mathfrak{P},i} \in O_L \otimes_{\mathbf{Z}} R$

$$\mathrm{F}\big(\mathfrak{e}_{\mathfrak{P},i-1} \cdot \eta\big) \in \big(O_L \underset{\mathbf{Z}}{\otimes} R\big)\mathfrak{e}_{\mathfrak{P},i} \cdot \eta.$$

7.12 Definition. *Define the modular form of weight* $\chi_{\mathfrak{P},i-1}^{p}\chi_{\mathfrak{P},i}^{-1}$ *over* k

$$h_{\mathfrak{P},i} \in \mathbf{M}\big(k, \chi_{\mathfrak{P},i-1}^{p}\chi_{\mathfrak{P},i}^{-1}\big)$$

by the rule

$$\mathrm{F}\big(\mathfrak{e}_{\mathfrak{P},i-1} \cdot \eta\big) \equiv h_{\mathfrak{P},i}\Big(\big(A, \iota, \lambda, \omega\big)\Big)\mathfrak{e}_{\mathfrak{P},i} \cdot \eta \qquad \bmod \mathfrak{P}.$$

See 4.5 for the notation.

The modular forms $\big\{h_{\mathfrak{P},i}\big\}_{\mathfrak{P},i}$ *are called the generalized or partial Hasse invariants. If* $\psi \in \mathbb{X}_k$ *is a weight of the form* $\psi = \prod_{\mathfrak{P},i}\big(\chi_{\mathfrak{P},i-1}^{p}\chi_{\mathfrak{P},i}^{-1}\big)^{a_{\mathfrak{P},i}}$*, with* $a_{\mathfrak{P},i} \in \mathbf{N}$*, define the modular form of weight* ψ

$$h_\psi := \prod_{\mathfrak{P},i} h_{\mathfrak{P},i}^{a_{\mathfrak{P},i}}.$$

Define $h := h_{\mathbf{Nm}^{p-1}}$ *as the Hasse invariant.*

7.13 Remark. It is easy to verify that all the conditions given in the definition 5.1 of modular forms of a given weight are satisfied.

7.14 Proposition. 1. *The q-expansion of the partial Hasse invariant $h_{\mathfrak{P},i}$ at any $\mathfrak{J}$-polarized unramified cusp defined over $\mathbf{F}_p$ is 1.*

2. *The modular form h coincides, up to a sign, with the determinant of the Hasse-Witt matrix.*

3. *The subgroup of $\mathbb{X}_k$ spanned by the weights of the partial Hasse invariants coincides with the subgroup $\mathbb{X}_k(1)$ of the elements $\chi \in \mathbb{X}_k$ such that $\chi\big((O_L/pO_L)^*\big) \equiv 1$ in k.*

Proof: Part (1) is proved via an explicit computation using Tate objects; c.f. [Go1, Thm. 2.1(2)]. Part (2) is clear. We next prove part (3). The set of basic characters $\{\chi_{\mathfrak{P},i} : \mathfrak{P}|p,\ 1 \le i \le f_{\mathfrak{P}}\}$ forms a basis for the lattice $\mathbb{X}_k$. The subgroup H of $\mathbb{X}_k$ spanned by the weights $\Big\{\chi_{\mathfrak{P},i-1}^{p}\chi_{\mathfrak{P},i}^{-1} : \mathfrak{P}|p,\ 1 \le i \le f_{\mathfrak{P}}\Big\}$ of the partial Hasse invariants is contained in $\mathbb{X}_k(1)$. We conclude by remarking that $\mathbb{X}_k/\mathbb{X}_k(1) \cong \prod_{\mathfrak{P}|p} \mathrm{Hom}_{\mathrm{Gr}}\big(\mathrm{k}_{\mathfrak{P}}^*, k^*\big)$ has cardinality $\prod_{\mathfrak{P}|p}\big(p^{f_{\mathfrak{P}}} - 1\big)$, the same as $\mathbb{X}_k/H$.

7.15 *Examples.* Assume that the prime p is inert in L. Let $\chi_1, \ldots, \chi_g$ be the basic characters of $\mathcal{G}_k$ ordered so that $\sigma \circ \chi_i = \chi_{i+1}$. Then we get precisely g partial Hasse invariants $h_1, \ldots, h_g$ of weights $\chi_g^p\chi_1^{-1}, \chi_1^p\chi_2^{-1}, \ldots, \chi_{g-1}^p\chi_g^{-1}$. In this case

$$\big(\chi_g^p\chi_1^{-1}\big)^{\mathbf{Z}} \times \cdots \times \big(\chi_{g-1}^p\chi_g^{-1}\big)^{\mathbf{Z}} = \mathbb{X}_k(1) \hookrightarrow \mathbb{X}_k = \chi_1^{\mathbf{Z}} \times \cdots \times \chi_g^{\mathbf{Z}}.$$

Assume that $p = \mathfrak{P}^g$ is totally ramified in L. Let Ψ be the unique basic character of $\mathcal{G}_k$. Then we get a unique partial Hasse invariant $h_{\Psi^{p-1}}$ which is a g-th root of the Hasse invariant and is of weight Ψ^{p-1}. We have

$$\left(\Psi^{p-1}\right)^{\mathbf{Z}} = \mathbb{X}_k(1) \hookrightarrow \mathbb{X}_k = \Psi^{\mathbf{Z}}.$$

7.16 *Second construction of the Hasse invariants: the definition by descent theory.* From 7.8 we deduce the following result:

7.17 Corollary. *The section $a(\chi)$ of the sheaf $\phi^*(\mathcal{L}_\chi)$ on $\mathfrak{M}(R/\mathfrak{m}^m, \mu_{p^n N})^{\mathrm{Kum}}$ descends to a section of $\mathcal{L}_\chi$ on $\mathfrak{M}(R/\mathfrak{m}^m, \mu_N)^{\mathrm{ord}}$ if and only if*

$$\chi(\alpha) = 1 \in (R/\mathfrak{m}^m)^* \qquad \text{for all } \alpha \in \left(O_L/p^n O_L\right)^*.$$

7.18 Corollary. *Let $\mathfrak{P}$ be a prime of O_L over p and let $1 \leq i \leq f_\mathfrak{P}$. The section $a(\chi^p_{\mathfrak{P},i-1}\chi^{-1}_{\mathfrak{P},i})$ descends to a section of the invertible sheaf $\mathcal{L}_{\chi^p_{\mathfrak{P},i-1}\chi^{-1}_{\mathfrak{P},i}}$ over $\mathfrak{M}(k, \mu_N)^{\mathrm{ord}}$. It coincides with the restriction of the partial Hasse invariant $h_{\mathfrak{P},i}$ defined in 7.12.*

Proof: The two modular forms have the same weight and the same q-expansion at any $\mathbf{F}_p$-rational cusp. Hence, they must be equal.

7.19 Definition. *Let*

$$R_{m,n} := \Gamma\left(\mathfrak{M}(R/\mathfrak{m}^m, \mu_{p^n N})^{\mathrm{Kum}}, O_{\mathfrak{M}\left(R/\mathfrak{m}^m, \mu_{p^n N}\right)^{\mathrm{Kum}}}\right).$$

Define the map

$$r\colon \bigoplus_{\chi \in \mathbb{X}^U_{R/\mathfrak{m}^m}} \mathbf{M}(R/\mathfrak{m}^m, \mu_N, \chi) \longrightarrow R_{m,n}$$

by the formula

$$f = \bigoplus_\chi f_\chi \longmapsto r(f) := \sum_\chi \frac{\phi^*(f_\chi)}{a(\chi)}.$$

Let $\alpha \in \left(O_L \otimes_{\mathbf{Z}} \mathbf{W}(k)\right)^*$. If f is a $\mathfrak{I}$-polarized modular form on $\mathfrak{M}(R/\mathfrak{m}^m, \mu_N)$ define

$$[\alpha]f(A, \iota, \lambda, \omega) := f(A, \iota, \lambda, \alpha^{-1}\omega).$$

This provides a graded action of $\left(O_L \otimes_{\mathbf{Z}} R\right)^*$ on $\oplus_{\chi \in \mathbb{X}^U_{R/\mathfrak{m}^m}} \mathbf{M}(R/\mathfrak{m}^m, \mu_N, \chi)$.

On the other hand the action of $G_{m,n}$, the Galois group of $\mathfrak{M}(R/\mathfrak{m}^m, \mu_{p^n N})^{\mathrm{Kum}} \to \mathfrak{M}(R/\mathfrak{m}^m, \mu_N)^{\mathrm{ord}}$, and the canonical projection $\left(O_L \otimes_{\mathbf{Z}} \mathbf{Z}/p^n\mathbf{Z}\right)^* \to G_{m,n}$, defined in 7.2, induce an action of $\left(O_L \otimes_{\mathbf{Z}} R\right)^*$ on $R_{m,n}$.

7.20 Proposition. *The map r defined in 7.19 has the following properties:*

1. *it is $\left(O_L \otimes_{\mathbf{Z}} R\right)^*$-equivariant;*

2. *let* $(\mathfrak{A},\mathfrak{B},\varepsilon,\mathrm{j}_\varepsilon)$ *be a* $\mathfrak{I}$*-polarized unramified cusp over* $R/\mathfrak{m}^m$ *of level* μ_{p^nN}*; see 6.4–6.5. The following diagram is commutative*

$$\begin{array}{ccc} \oplus_{\chi\in\mathbb{X}^U_{R/\mathfrak{m}^m}} \mathbf{M}\big(R/\mathfrak{m}^m,\mu_N,\chi\big) & \stackrel{r}{\longrightarrow} & R_{m,n} \\ & \searrow & \downarrow \\ & & R/\mathfrak{m}^m\otimes_{\mathbf{Z}}\mathbf{Z}\big((\mathfrak{A},\mathfrak{B},\sigma_\beta)\big). \end{array}$$

The notation is the following. The ring $\mathbf{Z}\big((\mathfrak{A},\mathfrak{B},\sigma_\beta)\big)$ *is the base over which* $\mathbf{Tate}(\mathfrak{A},\mathfrak{B})_{\sigma_\beta}$ *lives by 6.3, the vertical map is the unique one so that the pull-back of the universal abelian scheme over* $\mathfrak{M}\big(R/\mathfrak{m}^m,\mu_{p^nN}\big)$ *is* $\mathbf{Tate}(\mathfrak{A},\mathfrak{B})_{\sigma_\beta}$ *and the diagonal map is the* q*-expansion map at the cusp* $(\mathfrak{A},\mathfrak{B},\varepsilon,\mathrm{j}_\varepsilon)$ *defined in 6.6.*

Proof: Claim (1) follows from 7.8. Claim (2) follows from 6.6 and 7.9.

7.21 Corollary. *The kernel of the* q*-expansion map is independent of the cusp.*

7.22 Theorem. *Assume that* $m=n=1$*. In particular,* $R=k$*. Then*

1) *the kernel of* r *consists of the ideal* $\mathcal{I}$

$$\mathcal{I}:=\langle h_{\mathfrak{P},i}-1\,:\,\mathfrak{P}|p,\,1\le i\le f_{\mathfrak{P}}\rangle,$$

where the $h_{\mathfrak{P},i}$*'s are the partial Hasse invariants defined in 7.12;*

2) *the map* r *is surjective. In particular, the ring* $R_{1,1}$ *is canonically isomorphic to the ring of* $\mathfrak{I}$*-polarized modular forms* $\oplus_{\chi\in\mathbb{X}_k}\mathbf{M}\big(R/\mathfrak{m}^1,\mu_N,\chi\big)$ *modulo the ideal defined by the kernel of the* q*-expansion map.*

Proof: We prove (1). The fact that the q-expansion map is zero on the ideal $\mathcal{I}$ follows from 7.14. Let

$$f=\sum_\chi f_\chi\in\bigoplus_{\chi\in\mathbb{X}_k}\mathbf{M}\big(k,\mu_N,\chi\big)$$

be the sum of non-zero $\mathfrak{I}$-polarized modular forms f_χ of weight χ such that $r(f)=0$. Thanks to part (3) of 7.14, multiplying the modular forms f_χ's by suitable powers of the partial Hasse invariants, we can assume that the set of weights $\{\chi\}$ appearing in the decomposition of f does not contain two distinct elements defining the same character $\big(O_L/pO_L\big)^*\to k^*$. Consider the function $r(f)=\sum_\chi\phi^*\big(f_\chi\big)/a(\chi)$. It is the constant function 0 by assumption. By (2) of 7.20 the group $\big(O_L\otimes_{\mathbf{Z}}k\big)^*$ acts on $\phi^*\big(f_\chi\big)/a(\chi)$ via the character χ. Hence, each $\phi^*\big(f_\chi\big)/a(\chi)$ is zero i. e., $f_\chi=0$. We conclude that $f=0$.

We prove claim (2). The Galois group G of $\mathfrak{M}\big(k,\mu_{pN}\big)^{\mathrm{Kum}}\to\mathfrak{M}\big(k,\mu_N\big)^{\mathrm{ord}}$ is isomorphic to $\prod_{\mathfrak{P}|p}\mathrm{k}^*_{\mathfrak{P}}$ and, hence, has order prime to p. For every

$$\chi\in\mathbb{X}_k/\mathbb{X}_k(1)\cong\mathrm{Hom}_{\mathrm{Gr}}\big((O_L/pO_L)^*,k^*\big),$$

let

$$R^\chi_{1,1}:=\{b\in R_{1,1}|g\cdot b=\chi(g)\,b\quad\forall g\in G\}.$$

By Kummer theory we get a direct decomposition

$$R_{1,1}=\bigoplus_{\chi\in\mathbb{X}_k/\mathbb{X}_k(1)}R^\chi_{1,1}$$

into $R_{1,0}$-modules of rank 1. If $\chi \in \mathbb{X}_k$ and $b \in R^{\chi}_{1,1}$, then $b \cdot a(\chi)$ is a modular form of weight χ on $\mathfrak{M}(k, \mu_{pN})^{\text{Kum}}$. By 7.8 it descends to a $\mathfrak{I}$-polarized modular form of weight χ on $\mathfrak{M}(k, \mu_N)^{\text{ord}}$. Multiplying it by a suitable power h^s of the Hasse invariant, defined in 7.12, we may assume that $b \cdot a(\chi) h^s$ extends to a modular form f of weight $\chi \cdot \mathbf{Nm}^{s(p-1)}$ on $\mathfrak{M}(k, \mu_N)^{\text{R}}$. By construction, $r(f) = b$.

7.23 Corollary. *The ring* $\left(\oplus_{\chi \in \mathbb{X}_k} \mathbf{M}(R/\mathfrak{m}, \mu_N, \chi)\right)/\mathcal{I}$ *is a domain, finitely generated over* $R/\mathfrak{m}$.

Proof: Use the Theorem and 3.3.

7.24 *Exotic modular forms.* We prove the existence of $\mathfrak{I}$-polarized modular forms of weight χ over artinian bases for non-universal χ. A fortiori such modular forms can not be lifted to characteristic 0. We first need the following

7.25 Lemma. *Suppose that p is ramified. Let k be a finite field and let $m > 0$ be an integer. There exist infinitely many characters $\chi \in \mathbb{X}_{\mathbf{W}_{m+1}(k)}$ such that*
1) *χ is non-trivial and is not a universal character;*
2) *$\chi((O_L/p^{m+1}O_L)^*) = 1$ in $\mathbf{W}_{m+1}(k)$.*

Proof: We use the notation and the results of 4.8. Let $\chi \in \text{Ker}(\alpha)$ be a non-trivial character of $\mathcal{G}_{\mathbf{W}_{m+1}(k)}$. Let $\tilde{\chi}\colon \mathcal{G}_k \to \mathbf{G}_{a,k}$ be the associated non-trivial homomorphism of group schemes. Denote by σ the absolute Frobenius on k. Since k is a finite field, there exists a positive integer s such that k is killed by $\sigma^s - \text{Id}$. Let $b \in \mathcal{G}(\mathbf{W}_{m+1}(k))$ (e.g. $b \in \mathcal{G}(\mathbf{Z}/p^{m+1}\mathbf{Z}) = (O_L/p^{m+1}\mathbf{Z})^*$) and let $\bar{b} \in \mathcal{G}_k(k)$ be the reduction of b modulo p. Then $\chi(b) = 1$ if and only if $\tilde{\chi}(\bar{b}) = 0$. In particular, the character $\chi' \in \text{Ker}(\alpha)$ associated to $(\sigma^s - \text{Id}) \circ \tilde{\chi}$ is non-trivial and kills $\bar{b}$. Therefore, χ' satisfies the requirements of the lemma.

7.26 *Construction.* Let χ be as in 7.25. We have constructed in 7.8 a section $a(\chi)$ of $\phi^*(\mathcal{L}_\chi)$. By the properties of χ and using 7.8, it descends to a section of $\mathcal{L}_\chi$ over $\mathfrak{M}(\mathbf{W}_{m+1}(k), \mu_N)^{\text{ord}}$. Let h be the Hasse invariant defined on $\mathfrak{M}(k, \mu_N)^{\text{R}}$ in 7.14. We shall prove in 11.9 that there exists an integer t such that h^t lifts to a modular form $\hbar$ over $\mathfrak{M}(\mathbf{W}_{m+1}(k), \mu_N)^{\text{R}}$. It has the property that it vanishes exactly on the complement of the ordinary locus. In particular, there exists an integer s such that $a(\chi)\hbar^s$ extends to a global section of $\mathcal{L}_{\chi_1}$ over $\mathfrak{M}(\mathbf{W}_{m+1}(k), \mu_N)^{\text{R}}$, where $\chi_1 = \chi \cdot \mathbf{Nm}^{st(p-1)}$. Note that χ_1 is non-trivial and is not a universal character.

8 Reduceness of the partial Hasse invariants.

We prove in this section, using local deformation theory of displays, that the divisor of a partial Hasse invariant is reduced. This is an analogue of Igusa's theorem that states that the zeroes of the supersingular polynomial are simple. This result is the basis for establishing a notion of filtration for Hilbert modular forms of any (not necessarily parallel) weight, and for later computations regarding the operators U, V and $\Theta_{\mathfrak{P},i}$.

The following sections deal with the deformation theory of abelian varieties with real multiplication satisfying Rapoport's condition using Zink's theory of displays [Zi] (many of the results we are using can be found in the introduction to this paper). For the general case, and for extensive discussion of the relation between this deformation theory and the crystalline theory, see [AG].

8.1 *Deformation theory of abelian varieties with RM over the Rapoport locus.* Let $(A_0, \iota_0, \lambda_0)$ be a $\mathfrak{J}$-polarized abelian variety with RM by O_L over a perfect field k of characteristic p satisfying condition (R) defined in 3.5. Let $(A_0[p^\infty], \iota_0)$ be the associated p-divisible group with O_L-action. Let

$$(\mathrm{P}_0, \mathrm{Q}_0, \mathrm{F}_0, \mathrm{V}_0^{-1})$$

be the $3n$-display in the sense of [Zi] associated to $A_0[p^\infty]$. Note that λ_0 and the choices in 7.1 induce a polarization on A_0, and hence on $A_0[p^\infty]$, of degree prime to p. Let

$$\langle -, - \rangle_0 \colon \mathrm{P}_0 \times \mathrm{P}_0 \longrightarrow \mathbf{W}(k)$$

be the associated non-degenerate pairing of $3n$-displays as defined in [Zi]. The action of O_L on $A_0[p^\infty]$ induces an action of O_L on P_0 and Q_0 such that F_0 and V_0^{-1} are O_L-equivariant. Moreover,

$$\langle l\gamma, \delta \rangle_0 = \langle \gamma, l\delta \rangle_0$$

for all $l \in O_L \otimes_{\mathbf{Z}} \mathbf{W}(k)$ and for all γ and δ in P_0.

Note that the O_L-action on $A_0[p^\infty]$ induces a decomposition over all primes $\mathfrak{P}$ of O_L over p:

$$A_0[p^\infty] = \prod_{\mathfrak{P}|p} A_0[\mathfrak{P}^\infty].$$

Analogously the O_L-action on $(\mathrm{P}_0, \mathrm{Q}_0, \mathrm{F}_0, \mathrm{V}_0^{-1})$ and the O_L-linearity of F_0 and V_0^{-1} induce a decomposition

$$(\mathrm{P}_0, \mathrm{Q}_0, \mathrm{F}_0, \mathrm{V}_0^{-1}) = \prod_{\mathfrak{P}|p} (O_{L,\mathfrak{P}} \underset{O_L}{\otimes} \mathrm{P}_0, O_{L,\mathfrak{P}} \underset{O_L}{\otimes} \mathrm{Q}_0, \mathrm{F}_0, \mathrm{V}_0^{-1}).$$

For each prime $\mathfrak{P}$ the display $(O_{L,\mathfrak{P}} \otimes_{O_L} \mathrm{P}_0, O_{L,\mathfrak{P}} \otimes_{O_L} \mathrm{Q}_0, \mathrm{F}_0, \mathrm{V}_0^{-1})$ is associated to the p-divisible group $A_0[\mathfrak{P}^\infty]$.

8.2 Proposition. *The $O_L \otimes_{\mathbf{Z}} \mathbf{W}(k)$-module P_0 is free of rank 2. There exist α and β in P_0 such that*

$$\mathrm{P}_0 = (O_L \underset{\mathbf{Z}}{\otimes} \mathbf{W}(k))\alpha \oplus (O_L \underset{\mathbf{Z}}{\otimes} \mathbf{W}(k))\beta$$

and

$$\mathrm{Q}_0 = (O_L \underset{\mathbf{Z}}{\otimes} \mathbf{W}(k))p\alpha \oplus (O_L \underset{\mathbf{Z}}{\otimes} \mathbf{W}(k))\beta.$$

Moreover,

$$\mathfrak{T}_0 := (O_L \underset{\mathbf{Z}}{\otimes} \mathbf{W}(k))\alpha \qquad \text{and} \qquad \mathfrak{L}_0 := (O_L \underset{\mathbf{Z}}{\otimes} \mathbf{W}(k))\beta$$

are totally isotropic with respect to $\langle -, - \rangle_0$.

Proof: Noting that $\mathrm{P}_0 \overset{\sim}{\longrightarrow} \mathrm{H}_{1,\mathrm{crys}}(A_0/\mathbf{W}(k))$ and using [Ra, Lem. 1.3], we deduce that P_0 is a free $O_L \otimes_{\mathbf{Z}} \mathbf{W}(k)$-module of rank 2. The image $\bar{\mathrm{Q}}_0$ of Q_0 in $\bar{\mathrm{P}}_0 :=$

$\mathrm{P}_0/p\mathrm{P}_0$ is isomorphic to $\mathrm{H}^0\big(A_0^\vee, \Omega^1_{A_0^\vee/k}\big)$. Via λ_0 it is isomorphic to $\mathrm{H}^0\big(A_0, \Omega^1_{A_0/k}\big)$. In particular, since condition (R) holds, it is a free $O_L \otimes_{\mathbf{Z}} k$ module of rank 1. Let β be an element of Q_0 generating $\bar{\mathrm{Q}}_0$ as $O_L \otimes_{\mathbf{Z}} k$-module. The quotient $\bar{P}_0/\bar{\mathrm{Q}}_0$ is isomorphic to $\mathrm{Hom}_k\big(\mathrm{H}^0(A_0, \Omega^1_{A_0/k}), k\big)$ and, hence, it is a free $O_L \otimes_{\mathbf{Z}} k$-module of rank 1. Let $\alpha \in \mathrm{P}_0$ be an element generating $\bar{P}_0/\bar{\mathrm{Q}}_0$ as $O_L \otimes_{\mathbf{Z}} k$-module. For all $\gamma \in \mathrm{P}_0$

$$\langle l\gamma, \gamma\rangle_0 = \langle \gamma, l\gamma\rangle_0 = -\langle l\gamma, \gamma\rangle_0.$$

We conclude that $\langle l\gamma, \gamma\rangle_0 = 0$ for all $l \in O_L \otimes_{\mathbf{Z}} \mathbf{W}(k)$ and all $\gamma \in \mathrm{P}_0$. Hence the conclusion.

8.3 *Notation.* For each prime $\mathfrak{P}$ of O_L dividing p and each integer $1 \le i \le f_{\mathfrak{P}}$, let $\mathfrak{e}_{\mathfrak{P},i}$ be the associated idempotent of $O_L \otimes_{\mathbf{Z}} \mathbf{W}(k)$ defined in 2.1 and let $\pi_{\mathfrak{P}} \in O_L$ be the generator of the ideal $\mathfrak{P}\, O_{L,\mathfrak{P}}$ chosen in 2.1. The elements

$$\mathfrak{e}_{\mathfrak{P},i}^{[1]} := \mathfrak{e}_{\mathfrak{P},i}, \quad \mathfrak{e}_{\mathfrak{P},i}^{[2]} := \pi_{\mathfrak{P}}\mathfrak{e}_{\mathfrak{P},i}, \quad \ldots \quad , \mathfrak{e}_{\mathfrak{P},i}^{[e_{\mathfrak{P}}]} := \pi_{\mathfrak{P}}^{e_{\mathfrak{P}}-1}\mathfrak{e}_{\mathfrak{P},i}$$

form a $\mathbf{W}(k)$-basis of the module $\big(O_L \otimes_{\mathbf{Z}} \mathbf{W}(k)\big) \cdot \mathfrak{e}_{\mathfrak{P},i}$. Let $\alpha \in \mathrm{P}_0$ be as in 8.2. Define $\alpha_{\mathfrak{P},i}^{[j]} \in \mathrm{P}_0$ by

$$\alpha_{\mathfrak{P},i}^{[j]} := \mathfrak{e}_{\mathfrak{P},i}^{[j]}\alpha = \pi_{\mathfrak{P}}^{j-1}\mathfrak{e}_{\mathfrak{P},i}\alpha$$

for every prime $\mathfrak{P}$ over p and for all $1 \le i \le f_{\mathfrak{P}}$ and $1 \le j \le e_{\mathfrak{P}}$. Denote by

$$\beta_{\mathfrak{P},i}^{[j]}$$

the element of $\mathfrak{L}_0$ such that

$$\langle \alpha_{\mathfrak{P},i}^{[j]}, \beta_{\mathfrak{P},i}^{[j]}\rangle_0 = 1, \quad \text{and} \quad \langle \alpha_{\mathfrak{Q},s}^{[t]}, \beta_{\mathfrak{P},i}^{[j]}\rangle_0 = 0$$

if $\alpha_{\mathfrak{Q},s}^{[t]} \neq \alpha_{\mathfrak{P},i}^{[j]}$.

For every prime $\mathfrak{P}$ dividing p, define a total ordering

$$\alpha_{\mathfrak{P},i_1}^{[j_1]} < \alpha_{\mathfrak{P},i_2}^{[j_2]} \qquad \begin{cases} \text{if } i_1 < i_2; \\ \text{if } i_1 = i_2 \text{ and } j_1 < j_2. \end{cases}$$

Analogously for the $\beta_{\mathfrak{P},i}^{[j]}$'s. The elements $\left\{\alpha_{\mathfrak{P},i}^{[j]}\right\}_{i,j}$ form an ordered $\mathbf{W}(k)$-basis of the module $O_{L,\mathfrak{P}} \otimes_{O_L} \mathfrak{T}_0$. Analogously the elements $\left\{\beta_{\mathfrak{P},i}^{[j]}\right\}_{i,j}$ form an ordered $\mathbf{W}(k)$-basis of $O_{L,\mathfrak{P}} \otimes_{O_L} \mathfrak{L}_0$. By construction

$$\mathcal{B}_{\mathfrak{P}} := \left\{\alpha_{\mathfrak{P},i}^{[j]}, \beta_{\mathfrak{P},i}^{[j]}\right\}_{i,j}$$

is a symplectic basis for $\big(O_{L,\mathfrak{P}} \otimes_{O_L} \mathrm{P}_0, \langle\_,\_\rangle_0\big)$ as a $\mathbf{W}(k)$-module.

Let

$$\begin{pmatrix} A_{\mathfrak{P}} & B_{\mathfrak{P}} \\ C_{\mathfrak{P}} & D_{\mathfrak{P}} \end{pmatrix}$$

be the matrix of $\mathrm{F}_0 \oplus \mathrm{V}_0^{-1}$ on $(O_{L,\mathfrak{P}} \otimes_{O_L} \mathfrak{T}_0) \oplus (O_{L,\mathfrak{P}} \otimes_{O_L} \mathfrak{L}_0)$ with respect to the given basis. Define $c_{\mathfrak{P},i}^{[1]}$ and $a_{\mathfrak{P},i}^{[1]}$ in $\mathbf{W}(k)$ by

$$C_{\mathfrak{P}}\left(\alpha_{\mathfrak{P},i-1}^{[1]}\right) = c_{\mathfrak{P},i}^{[1]}\beta_{\mathfrak{P},i}^{[1]} + \cdots \qquad \text{and} \qquad A_{\mathfrak{P}}\left(\alpha_{\mathfrak{P},i-1}^{[1]}\right) = a_{\mathfrak{P},i}^{[1]}\alpha_{\mathfrak{P},i}^{[1]} + \cdots.$$

8.4 Remark. For primes $\mathfrak{P}$, $\mathfrak{Q}$ of O_L over p, we have by construction

$$\langle \alpha_{\mathfrak{Q},s}^{[t-1]}, \pi_{\mathfrak{Q}} \beta_{\mathfrak{P},i}^{[j]} \rangle_0 = \langle \pi_{\mathfrak{Q}} \alpha_{\mathfrak{Q},s}^{[t-1]}, \beta_{\mathfrak{P},i}^{[j]} \rangle_0 = \begin{cases} \langle \alpha_{\mathfrak{Q},s}^{[t]}, \beta_{\mathfrak{P},i}^{[j]} \rangle_0 & \text{if } 2 \leq t \leq e_{\mathfrak{Q}} \\ \langle \gamma, \beta_{\mathfrak{P},i}^{[j]} \rangle_0 & \gamma \in p\,\mathrm{P}_0, \text{ if } t = e_{\mathfrak{Q}} + 1. \end{cases}$$

Hence, for any $2 \leq j \leq e_{\mathfrak{P}}$, we have $\pi_{\mathfrak{P}} \beta_{\mathfrak{P},i}^{[j]} = \beta_{\mathfrak{P},i}^{[j-1]}$ modulo $p\,\mathrm{P}_0$.

8.5 Lemma. *Let $\mathfrak{P}$ be a prime over p;*

1. *the $2\,(f_{\mathfrak{P}} e_{\mathfrak{P}}) \times (f_{\mathfrak{P}} e_{\mathfrak{P}})$ matrix*

$$\begin{pmatrix} A_{\mathfrak{P}} \\ C_{\mathfrak{P}} \end{pmatrix}$$

has rank g;

2. *for any i, j as above we have*

$$A_{\mathfrak{P}}\big(\alpha_{\mathfrak{P},i}^{[j]}\big) \in \bigoplus_{s=1}^{e_{\mathfrak{P}}} \mathbf{W}(k)\, \alpha_{\mathfrak{P},i+1}^{[s]} \qquad \text{and} \qquad C_{\mathfrak{P}}\big(\alpha_{\mathfrak{P},i}^{[j]}\big) \in \bigoplus_{s=1}^{e_{\mathfrak{P}}} \mathbf{W}(k)\, \beta_{\mathfrak{P},i+1}^{[s]}.$$

Analogously,

$$B_{\mathfrak{P}}\big(\beta_{\mathfrak{P},i}^{[j]}\big) \in \bigoplus_{s=1}^{e_{\mathfrak{P}}} \mathbf{W}(k)\, \alpha_{\mathfrak{P},i+1}^{[s]} \qquad \text{and} \qquad D_{\mathfrak{P}}\big(\beta_{\mathfrak{P},i}^{[j]}\big) \in \bigoplus_{s=1}^{e_{\mathfrak{P}}} \mathbf{W}(k)\, \beta_{\mathfrak{P},i+1}^{[s]};$$

3. *the matrix $A_{\mathfrak{P}}$ is invertible if and only if $A_0[\mathfrak{P}^\infty]$ is ordinary.*

Proof: The first assertion follows since the map $\mathrm{F}_0 \oplus \mathrm{V}_0^{-1}$ is an isomorphism. The second assertion follows the σ-linearity of F_0 and of V_0^{-1} and from the definition of the elements $\alpha_{\mathfrak{P},i}^{[j]}$ and $\beta_{\mathfrak{P},i}^{[j]}$. Note that $A_0[\mathfrak{P}^\infty]$ is ordinary if and only if the reduction of F_0 on $(O_L/\mathfrak{P}) \otimes_{O_L} (\bar{\mathrm{P}}_0/\bar{\mathrm{Q}}_0) = \mathfrak{T}_0/\mathfrak{P}\mathfrak{T}_0$ is an isomorphism. This proves the last assertion.

8.6 Corollary. *Let $\mathfrak{P}$ be a prime over p. Then, either $A_0[\mathfrak{P}^\infty]$ is ordinary or it is connected.*

Proof: The $3n$-display defined by $A_0[\mathfrak{P}^\infty]$ is $\big(O_{L,\mathfrak{P}} \otimes_{O_L} \mathrm{P}_0, O_{L,\mathfrak{P}} \otimes_{O_L} \mathrm{Q}_0, \mathrm{F}_0, \mathrm{V}_0^{-1}\big)$. By 8.2, we have that $O_{L,\mathfrak{P}} \otimes_{O_L} \mathfrak{T}_0$ is free of rank 1 over $O_{L,\mathfrak{P}} \otimes_{\mathbf{Z}_p} \mathbf{W}(k)$. Consider the reduction $\bar{A}_{\mathfrak{P}}$ on $\mathfrak{T}_0/\mathfrak{P}\mathfrak{T}_0$ of the matrix $A_{\mathfrak{P}}$. By 8.5, Part (2), either $\bar{A}_{\mathfrak{P}}$ is invertible or nilpotent. The p-divisible group $A_0[\mathfrak{P}^\infty]$ is connected if and only if $A_0[\mathfrak{P}]$ is connected. This is equivalent to ask that $\bar{A}_{\mathfrak{P}}$ is nilpotent. We conclude by 8.5, Part (3).

8.7 Proposition. *Let $\mathfrak{P}$ be a prime over p. Assume that $A_0[\mathfrak{P}^\infty]$ has p-rank equal to 0 (equiv. is non-ordinary by 8.6). The universal equi-characteristic deformation space of $A_0[\mathfrak{P}^\infty]$, as a principally polarized p-divisible group, is $R_{\mathfrak{P}} := k[\![t_{a,b}]\!]_{1 \leq a,b \leq f_{\mathfrak{p}} e_{\mathfrak{P}}}$ with the relations $t_{a,b} = t_{b,a}$. The universal display, denoted by $\big(\mathrm{P}_{\mathfrak{P}}, \mathrm{Q}_{\mathfrak{P}}, \mathrm{F}_{\mathfrak{P}}, \mathrm{V}_{\mathfrak{P}}^{-1}\big)$, is given by*

$$\mathrm{P}_{\mathfrak{P}} := \Big(O_{L,\mathfrak{P}} \underset{O_L}{\otimes} \mathrm{P}_0\Big) \underset{\mathbf{W}(k)}{\otimes} \mathbf{W}(R_{\mathfrak{P}}), \quad \mathrm{Q}_{\mathfrak{P}} := \Big(O_{L,\mathfrak{P}} \underset{O_L}{\otimes} \mathrm{Q}_0\Big) \underset{\mathbf{W}(k)}{\otimes} \mathbf{W}(R_{\mathfrak{P}})$$

and

$$\mathfrak{L}_{\mathfrak{P}} := \Big(O_{L,\mathfrak{P}} \underset{O_L}{\otimes} \mathfrak{L}_0\Big) \underset{\mathbf{W}(k)}{\otimes} \mathbf{W}(R_{\mathfrak{P}}), \quad \mathfrak{T}_{\mathfrak{P}} := \Big(O_{L,\mathfrak{P}} \underset{O_L}{\otimes} \mathfrak{T}_0\Big) \underset{\mathbf{W}(k)}{\otimes} \mathbf{W}(R_{\mathfrak{P}}).$$

The matrix of $\mathrm{F}_{\mathfrak{P}} \oplus \mathrm{V}_{\mathfrak{P}}^{-1}$ *on* $\mathfrak{T}_{\mathfrak{P}} \oplus \mathfrak{L}_{\mathfrak{P}}$ *with respect to* $\mathcal{B}_P$ *is*

$$\begin{pmatrix} A_{\mathfrak{P}} + T_{\mathfrak{P}} C_{\mathfrak{P}} & B_{\mathfrak{P}} + T_{\mathfrak{P}} D_{\mathfrak{P}} \\ C_{\mathfrak{P}} & D_{\mathfrak{P}} \end{pmatrix},$$

where $T_{\mathfrak{P}}$ *is the symmetric matrix of Teichmüller lifts* $\left(w(t_{a,b})\right)_{1 \le a,b \le e_{\mathfrak{P}} f_{\mathfrak{P}}}$. *The pairing*

$$\langle -, - \rangle_{\mathfrak{P}},$$

defined extending $O_{L,\mathfrak{P}} \otimes_{O_L} \mathbf{W}(R_{\mathfrak{P}})$*-linearly the pairing* $\langle -, - \rangle_0$*, is a non-degenerate pairing of displays such that* $\mathfrak{L}_{\mathfrak{P}}$ *and* $\mathfrak{T}_{\mathfrak{P}}$ *are maximal isotropic submodules of* $\mathrm{P}_{\mathfrak{P}}$.

Proof: It follows form the assumption on the p-rank of $A_0[p^\infty]$ that $(\mathrm{P}_0, \mathrm{Q}_0, \mathrm{F}_0, \mathrm{V}_0^{-1})$ is a display. The theorem follows from [Zi, §2.2].

8.8 *The universal equi-characteristic deformation space of* (A_0, ι_0). By the Serre-Tate theorem the universal equi-characteristic deformation space $\mathrm{Spf}(R_\iota)$ of A_0 with the O_L-action coincides with the universal equi-characteristic deformation space $\mathrm{Spf}(R_\iota)$ of $A_0[p^\infty]$ with the O_L-action. Hence,

$$\mathrm{Spf}(R_\iota) = \prod_{\mathfrak{P}|p} \mathrm{Spf}(R_{\mathfrak{P},\iota}),$$

where for any prime $\mathfrak{P}$ we define $\mathrm{Spf}(R_{\mathfrak{P},\iota})$ as the universal deformation space of $A_0[\mathfrak{P}^\infty]$ with O_L-action.

Fix a prime $\mathfrak{P}$. If $A_0[\mathfrak{P}^\infty]$ is non-ordinary, then $\mathrm{Spf}(R_{\mathfrak{P},\iota})$ is the closed subscheme of $\mathrm{Spf}(R_{\mathfrak{P}})$ defined by the condition that $\mathrm{F}_{\mathfrak{P}}$ and $\mathrm{V}_{\mathfrak{P}}^{-1}$ commute with the O_L-action on $\mathrm{P}_{\mathfrak{P}}$ and $\mathrm{Q}_{\mathfrak{P}}$ induced by the O_L-structures of P_0 and Q_0. Since $\langle \mathrm{F}_{\mathfrak{P}}(x), \mathrm{V}_{\mathfrak{P}}^{-1}(y) \rangle = \langle x, y \rangle^\sigma$ for all x in $\mathrm{P}_{\mathfrak{P}}$ and all y in $\mathrm{Q}_{\mathfrak{P}}$, we deduce that $\mathrm{F}_{\mathfrak{P}} \oplus \mathrm{V}_{\mathfrak{P}}^{-1}$ is a symplectic isomorphism. Hence, $\mathrm{V}_{\mathfrak{P}}^{-1}$ is O_L-linear if and only if $\mathrm{F}_{\mathfrak{P}}$ is. This is equivalent to require that $\mathrm{F}_{\mathfrak{P}}$ restricted to $\mathfrak{T}_{\mathfrak{P}}$ i. e., $A_{\mathfrak{P}} + T_{\mathfrak{P}} C_{\mathfrak{P}}$, is equivariant with respect to the O_L-structure on $\mathfrak{T}_{\mathfrak{P}}$.

8.9 Lemma. *The conditions that the restriction of* $\mathrm{F}_{\mathfrak{P}}$ *to* $\mathfrak{T}_{\mathfrak{P}}$ *is* O_L*-equivariant are linear in the* $t_{a,b}$*'s and are equivalent to the conditions*

1. $\mathrm{F}_{\mathfrak{P}}(\alpha_{\mathfrak{P},i}^{[1]}) \in \mathbf{W}(R)\alpha_{\mathfrak{P},i+1}^{[1]} \oplus \ldots \oplus \mathbf{W}(R)\alpha_{\mathfrak{P},i+1}^{[e_{\mathfrak{P}}]}$ *for all* $1 \le i \le f_{\mathfrak{P}}$;
2. $\mathrm{F}_{\mathfrak{P}}(\alpha_{\mathfrak{P},i}^{[j]}) = \pi_{\mathfrak{P}}^{j-1} \mathrm{F}(\alpha_{\mathfrak{P},i}^{[1]})$ *for all* $1 \le i \le f_{\mathfrak{P}}$ *and all* $1 \le j \le e_{\mathfrak{P}}$.

Proof: Clear.

8.10 Definition. *For any prime* $\mathfrak{P}$ *and integers* $1 \le i \le f_{\mathfrak{P}}$ *and* $1 \le j \le e_{\mathfrak{P}}$ *let*

$$t_{\mathfrak{P},i}^{[j]} := t_{a,b}$$

with $a = (i-1)e_{\mathfrak{P}} + 1$ *and* $b = (i-1)e_{\mathfrak{P}} + 1 + (j-1)$.

8.11 Theorem. *Let* $\mathfrak{P}$ *be a prime over* p. *Assume that* $A_0[\mathfrak{P}^\infty]$ *is non-ordinary. Let* $\mathrm{Spf}(R_{\mathfrak{P},\iota})$ *be the universal equi-characteristic deformation space of* $A_0[\mathfrak{P}^\infty]$ *as* p*-divisible group with* O_L*-action. Then,* $R_{\mathfrak{P},\iota}$ *is a* $(f_{\mathfrak{P}} e_{\mathfrak{P}})$*-dimensional power series ring*

$$R_{\mathfrak{P},\iota} = k[\![t_{\mathfrak{P},i}^{[j]}]\!]_{i,j}.$$

The associated universal display $(\mathrm{P}_{\mathfrak{P}}, \mathrm{Q}_{\mathfrak{P}}, \mathrm{F}_{\mathfrak{P}}, \mathrm{V}_{\mathfrak{P}}^{-1})$ *is given by*

$$\mathrm{P}_{\mathfrak{P}} := \left(O_{L,\mathfrak{P}} \underset{O_L}{\otimes} \mathrm{P}_0\right) \underset{\mathbf{W}(k)}{\otimes} \mathbf{W}(R_{\mathfrak{P},\iota}), \quad \mathrm{Q}_{\mathfrak{P}} := \left(O_{L,\mathfrak{P}} \underset{O_L}{\otimes} \mathrm{Q}_0\right) \underset{\mathbf{W}(k)}{\otimes} \mathbf{W}(R_{\mathfrak{P},\iota})$$

and

$$\mathfrak{L}_{\mathfrak{P}} := \left(O_{L,\mathfrak{P}} \underset{O_L}{\otimes} \mathfrak{L}_0\right) \underset{\mathbf{W}(k)}{\otimes} \mathbf{W}(R_{\mathfrak{P},\iota}), \quad \mathfrak{T}_{\mathfrak{P}} := \left(O_{L,\mathfrak{P}} \underset{O_L}{\otimes} \mathfrak{T}_0\right) \underset{\mathbf{W}(k)}{\otimes} \mathbf{W}(R_{\mathfrak{P},\iota}).$$

The matrix of $\mathrm{F}_{\mathfrak{P}} \oplus \mathrm{V}_{\mathfrak{P}}^{-1}$ *on* $\mathfrak{T}_{\mathfrak{P}} \oplus \mathfrak{L}_{\mathfrak{P}}$ *with respect to* $\mathcal{B}_P$ *is*

$$\begin{pmatrix} A_{\mathfrak{P}} + T_{\mathfrak{P}} C_{\mathfrak{P}} & B_{\mathfrak{P}} + T_{\mathfrak{P}} D_{\mathfrak{P}} \\ C_{\mathfrak{P}} & D_{\mathfrak{P}} \end{pmatrix},$$

where $T_{\mathfrak{P}}$ *is the matrix determined by*

$$T_{\mathfrak{P}}\left(\alpha_{\mathfrak{P},i}^{[j]}\right) = \sum_{l=1}^{e_{\mathfrak{P}}} w\left(t_{\mathfrak{P},i}^{[l]}\right) \pi_{\mathfrak{P}}^{j-1} \alpha_{\mathfrak{P},i}^{[l]}$$

for all integers $1 \le i \le f_{\mathfrak{P}}$ *and* $1 \le j \le e_{\mathfrak{P}}$.

Proof: By 8.8 and 8.9, the formal scheme $\mathrm{Spf}(R_{\mathfrak{P},\iota})$ is the universal deformation space of $(O_{L,\mathfrak{P}} \otimes_{O_L} \mathrm{P}_0, O_{L,\mathfrak{P}} \otimes_{O_L} \mathrm{Q}_0, \mathrm{F}_0, \mathrm{V}_0^{-1})$ as display with O_L-action. By [Zi], it is also the universal deformation space of $A_0[\mathfrak{P}^\infty]$ with the O_L-action.

8.12 Corollary. *The notation is as above. The formal scheme* $\mathrm{Spf}(R_{\mathfrak{P},\iota})$ *is formally smooth of dimension* $f_{\mathfrak{P}} e_{\mathfrak{P}}$.

8.13 *Example: the inert case.* In this case p remains a prime ideal in O_L. We omit the subscripts $\mathfrak{P}$ and $[j]$ in the formulas above. The matrix $A + TC$ mod p of 8.11 is

$$\begin{pmatrix} 0 & 0 & \dots & 0 & \bar{a}_1 + t_1 \bar{c}_1 \\ \bar{a}_2 + t_2 \bar{c}_2 & 0 & \dots & 0 & 0 \\ 0 & \bar{a}_3 + t_3 \bar{c}_3 & \dots & 0 & 0 \\ \vdots & \vdots & \vdots & \vdots & \vdots \\ 0 & 0 & \dots & \bar{a}_g + t_g \bar{c}_g & 0 \end{pmatrix}.$$

8.14 *Example: the totally ramified case.* Suppose that there is only one prime over p which is totally ramified. In this case we omit the subscripts $\mathfrak{P}$ and i. We write the coefficients $\bar{a}_{\mathfrak{P},i}^{[j]}, \bar{c}_{\mathfrak{P},i}^{[j]}$ and the variables $t_{\mathfrak{P},i}^{[j]}$ as $\bar{a}_{[j]}, \bar{c}_{[j]}$ and $t_{[j]}$, respectively. The matrix $A + TC$ mod p of 8.11 is

$$\begin{pmatrix} \bar{a}_{[1]} + t_{[1]} \bar{c}_{[1]} & 0 & \dots & 0 \\ \bar{a}_{[2]} + t_{[2]} \bar{c}_{[1]} + t_{[1]} \bar{c}_{[2]} & \bar{a}_{[1]} + t_{[1]} \bar{c}_{[1]} & \dots & 0 \\ \vdots & \vdots & \vdots & \vdots \\ \bar{a}_{[g]} + \sum_{i=1}^{g} t_{[g-i+1]} \bar{c}_{[i]} & \bar{a}_{[g-1]} + \sum_{i=1}^{g-1} t_{[g-i]} \bar{c}_{[i]} & \dots & \bar{a}_{[1]} + t_{[1]} \bar{c}_{[1]} \end{pmatrix}.$$

8.15 *The Hasse-Witt matrix of the universal equi-characteristic deformation.* Let (A, ι, λ) be the universal equi-characteristic object over the universal deformation space of $(A_0, \iota_0, \lambda_0)$. Let

$$(A^{\mathfrak{P}}, \iota^{\mathfrak{P}}, \lambda^{\mathfrak{P}}) := (A, \iota, \lambda) \underset{\mathrm{Spf}(R_\iota)}{\times} \mathrm{Spf}(R_{\mathfrak{P},\iota});$$

the morphism $\mathrm{Spf}(R_{\mathfrak{P},\iota}) \to \mathrm{Spf}(R_\iota) = \prod_{\mathfrak{Q}|p} \mathrm{Spf}(R_{\mathfrak{Q},\iota})$ (see 8.8 for the last equality) is defined to be the identity on the factor $\mathrm{Spf}(R_{\mathfrak{P},\iota})$ and the map $\mathrm{Spf}(R_{\mathfrak{P},\iota}) \to \mathrm{Spf}(k) \to \mathrm{Spf}(R_{\mathfrak{Q},\iota})$ if $\mathfrak{Q} \neq \mathfrak{P}$. Let $I_{R_{\mathfrak{P},\iota}}$ be the kernel of the reduction map $\mathbf{W}(R_{\mathfrak{P},\iota}) \to R_{\mathfrak{P},\iota}$. By [Me], we get a canonical identification of $\mathrm{H}_{1,\mathrm{dR}}(A^{\mathfrak{P}}/R_{\mathfrak{P},\iota})$ with the Lie algebra of the universal vector extension of $A^{\mathfrak{P}}[p^\infty]$. By the definition of $R_{\mathfrak{P},\iota}$ and $A^{\mathfrak{P}}$ we have

$$A^{\mathfrak{P}}[p^\infty] \cong A^{\mathfrak{P}}[\mathfrak{P}^\infty] \times \prod_{\mathfrak{Q}\neq\mathfrak{P}} \left(A_0[\mathfrak{Q}^\infty] \underset{k}{\times} \mathrm{Spf}(R_{\mathfrak{P},\iota}) \right).$$

Hence, for any prime $\mathfrak{Q}$ over p different from $\mathfrak{P}$ we get a canonical isomorphism $(O_L/\mathfrak{Q}) \otimes_{O_L} \mathrm{H}_{1,\mathrm{dR}}(A^{\mathfrak{P}}/R_{\mathfrak{P},\iota}) \cong (\mathrm{P}_0/\mathfrak{Q}\mathrm{P}_0) \otimes_k \mathrm{R}_{\mathfrak{P},\iota}$. By 8.11 we have a canonical identification

$$(O_L/\mathfrak{P}) \underset{O_L}{\otimes} \mathrm{H}_{1,\mathrm{dR}}(A^{\mathfrak{P}}/R_{\mathfrak{P},\iota}) \cong \mathrm{P}_{\mathfrak{P}}/I_{R_{\mathfrak{P},\iota}}\mathrm{P}_{\mathfrak{P}}.$$

The exact sequence

$$0 \longrightarrow \mathrm{Hom}\Big(\mathrm{H}^1(A^{\mathfrak{P}}, O_{A^{\mathfrak{P}}}), R_{\mathfrak{P},\iota}\Big) \longrightarrow \mathrm{H}_{1,\mathrm{dR}}(A^{\mathfrak{P}}/R_{\mathfrak{P},\iota})$$
$$\longrightarrow \mathrm{Hom}\Big(\mathrm{H}^0(A^{\mathfrak{P}}, \Omega^1_{A^{\mathfrak{P}}/R_{\mathfrak{P},\iota}}), R_{\mathfrak{P},\iota}\Big) \longrightarrow 0,$$

tensored over O_L with $O_L/\mathfrak{P}$ (resp. $O_L/\mathfrak{Q}$ for $\mathfrak{Q} \neq \mathfrak{P}$), and the exact sequence

$$0 \longrightarrow \mathfrak{L}_{\mathfrak{P}}/I_{R_{\mathfrak{P},\iota}}\mathfrak{L}_{\mathfrak{P}} \longrightarrow \mathrm{P}_{\mathfrak{P}}/I_{R_{\mathfrak{P},\iota}}\mathrm{P}_{\mathfrak{P}} \longrightarrow \mathfrak{T}_{\mathfrak{P}}/I_{R_{\mathfrak{P},\iota}}\mathfrak{T} \longrightarrow 0$$

(resp.

$$0 \longrightarrow (\mathfrak{L}_0/\mathfrak{Q}\mathfrak{L}_0) \underset{k}{\otimes} \mathrm{R}_{\mathfrak{P},\iota} \longrightarrow (\mathrm{P}_0/\mathfrak{Q}\mathrm{P}_0) \underset{k}{\otimes} \mathrm{R}_{\mathfrak{P},\iota} \longrightarrow (\mathfrak{T}_0/\mathfrak{Q}\mathfrak{T}_0) \underset{k}{\otimes} \mathrm{R}_{\mathfrak{P},\iota} \longrightarrow 0)$$

are identified. Using $\lambda^{\mathfrak{P}}$ and the choices in 7.1, we obtain a polarization on A of degree prime to p. This induces perfect pairings between $\mathrm{H}^0(A^{\mathfrak{P}}, \Omega^1_{A^{\mathfrak{P}}/R_{\mathfrak{P},\iota}})$ and $\mathrm{H}^1(A^{\mathfrak{P}}, O_{A^{\mathfrak{P}}})$, and between $\mathfrak{L}/I_{R_\iota}\mathfrak{L}$ and $\mathfrak{T}_{\mathfrak{P}}/I_{R_{\mathfrak{P},\iota}}\mathfrak{T}_{\mathfrak{P}}$ (resp. $(\mathfrak{T}_0/\mathfrak{Q}\mathfrak{T}_0) \otimes_k \mathrm{R}_{\mathfrak{P},\iota}$ and $(\mathfrak{L}_0/\mathfrak{Q}\mathfrak{L}_0) \otimes_k \mathrm{R}_{\mathfrak{P},\iota}$), compatible with the identifications given above. Hence, we get canonical isomorphisms

$$\mathrm{H}^0(A^{\mathfrak{P}}, \Omega^1_{A^{\mathfrak{P}}/R_{\mathfrak{P},\iota}}) \xrightarrow{\sim} \mathfrak{L}_0 \underset{\mathbf{W}(k)}{\otimes} \mathrm{R}_{\mathfrak{P},\iota}$$

and

$$\mathrm{H}^1(A^{\mathfrak{P}}, O_{A^{\mathfrak{P}}}) \xrightarrow{\sim} \mathfrak{T}_0 \underset{\mathbf{W}(k)}{\otimes} \mathrm{R}_{\mathfrak{P},\iota}$$

so that Frobenius on the left hand side, induced by Frobenius on $O_{A^{\mathfrak{P}}}$, corresponds to Frobenius on the right hand side.

By 8.3, the k-vector space $\mathfrak{T}_0/p\mathfrak{T}_0$ (resp. $\mathfrak{L}_0/p\mathfrak{L}_0$) is endowed with k-generators $\bar{\alpha}^{[j]}_{\mathfrak{Q},i}$ (resp. $\bar{\beta}^{[j]}_{\mathfrak{Q},i}$) defined as the reduction of $\alpha^{[j]}_{\mathfrak{Q},i}$ (resp. $\beta^{[j]}_{\mathfrak{Q},i}$). Their images induce canonical $R_{\mathfrak{P},\iota}$-generators

$$\{\eta^{[j]}_{\mathfrak{Q},i}\}_{\mathfrak{Q},i,j} \subset \mathrm{H}^1(A^{\mathfrak{P}}, O_{A^{\mathfrak{P}}})$$

(resp.

$$\{\omega^{[j]}_{\mathfrak{Q},i}\}_{\mathfrak{Q},i,j} \subset \mathrm{H}^0(A^{\mathfrak{P}}, \Omega^1_{A^{\mathfrak{P}}/R_{\mathfrak{P},\iota}})).$$

We deduce the following

8.16 Lemma. *Let $\mathfrak{P}$ be a prime over p. Let $\{\mathfrak{Q}_1, \ldots, \mathfrak{Q}_d\}$ be the set of primes over p different from $\mathfrak{P}$. Assume that $A_0[\mathfrak{P}^\infty]$ is not ordinary. The Hasse-Witt matrix of $(A^{\mathfrak{P}}, \iota^{\mathfrak{P}}, \lambda^{\mathfrak{P}})$ with respect to the basis $\{\eta^{[j]}_{\mathfrak{Q},i}\}_{\mathfrak{Q},i,j}$ is canonically identified with the reduction of the matrix*

$$\begin{pmatrix} A_{\mathfrak{Q}_1} & 0 & 0 & 0 \\ 0 & \cdots & A_{\mathfrak{Q}_d} & 0 \\ 0 & \cdots & 0 & A_{\mathfrak{P}} + T_{\mathfrak{P}} C_{\mathfrak{P}} \end{pmatrix}$$

via the quotient map $\mathbf{W}(R_\iota) \to R_\iota$.

8.17 Corollary. *Let $\mathfrak{P}$ be a prime of O_L over p. Assume that $A_0[\mathfrak{P}^\infty]$ is not ordinary. Let i be an integer satisfying $1 \leq i \leq f_{\mathfrak{P}}$. With the notation of 7.12, we have*

$$h_{\mathfrak{P},i}(A, \iota, \lambda, \omega_\alpha) = \bar{a}^{[1]}_{\mathfrak{P},i} + \bar{c}^{[1]}_{\mathfrak{P},i} t^{[1]}_{\mathfrak{P},i}.$$

The elements $\bar{a}^{[1]}_{\mathfrak{P},i}$ and $\bar{c}^{[1]}_{\mathfrak{P},i}$ are the reduction mod p of the element $a^{[1]}_{\mathfrak{P},i}$ and $c^{[1]}_{\mathfrak{P},i}$ defined in 8.3. If $\bar{a}^{[1]}_{\mathfrak{P},i} = 0$, then $\bar{c}^{[1]}_{\mathfrak{P},i}$ is invertible by 8.5.

8.18 Corollary. *The zero locus $W_{\mathfrak{P},i}$ of the partial Hasse invariant $h_{\mathfrak{P},i}$ is a reduced, non-singular divisor. In particular, it is locally irreducible.*

The divisor of the Hasse-invariant is equal to $\sum_{\mathfrak{P},i} e_{\mathfrak{P}} W_{\mathfrak{P},i}$ and the $W_{\mathfrak{P},i}$ are normal crossing divisors.

Note that this corollary, as all other results in this section, holds only over the Rapoport locus; the closure of $W_{\mathfrak{P},i}$ can be singular and even locally reducible.

8.19 Theorem. *Let $f \in \mathbf{M}(k, \mu_N, \chi)$ be a $\mathfrak{I}$-polarized modular form over k of weight χ. There is a unique $\mathfrak{I}$-polarized modular form g over k having the same q-expansion as f at some (hence, any) cusp and such that if g' is a $\mathfrak{I}$-polarized modular form over k with the same q-expansion of f, then there exist non-negative integers $b_{\mathfrak{P},i}$ for each prime $\mathfrak{P}$ of O_L over p and each integer $1 \leq i \leq f_{\mathfrak{P}}$ such that*

$$g' = g \prod_{\mathfrak{P},i} h^{b_{\mathfrak{P},i}}_{\mathfrak{P},i};$$

see 7.12 for the definition of the partial Hasse invariants $h_{\mathfrak{P},i}$.

Proof: Define g to be a $\mathfrak{I}$-polarized modular form satisfying

$$f = g \prod_{\mathfrak{P},i} h^{a_{\mathfrak{P},i}}_{\mathfrak{P},i},$$

where the $a_{\mathfrak{P},i}$ are chosen maximal non-negative so that g is a holomorphic modular form. The modular form g is unique with this property by the two previous corollaries. By 7.14, the modular form g has the same q-expansion of f. It satisfies the requirement of the Theorem by 7.22.

8.20 *Filtrations on modular forms.* The notation is as above. Define the *filtration* of f, denoted by

$$\Phi(f),$$

to be the weight of the unique $\mathfrak{I}$-polarized modular form g with the properties described in the Theorem. Since the weight of $h_{\mathfrak{P},i}$ is $\chi_{\mathfrak{P},i}^{p}\chi_{\mathfrak{P},i}^{-1}$, we have

$$\chi = \Phi(f)\prod_{\mathfrak{P},i}\left(\chi_{\mathfrak{P},i-1}^{p}\chi_{\mathfrak{P},i}^{-1}\right)^{a_{\mathfrak{P},i}}$$

for suitable non-negative integers $a_{\mathfrak{P},i}$.

9 A compactification of $\mathfrak{M}(k,\mu_{pN})^{\mathrm{Kum}}$.

Fix a field k of characteristic p containing all the finite fields $k_{\mathfrak{P}}$; see 7.1. Let $N \geq 4$ be an integer prime to p. In this section we construct a compactification of $\mathfrak{M}(k,\mu_{pN})^{\mathrm{Kum}}$, which is well suited for the study of the arithmetic of modular forms. The compactification is normal and it is explicit, in the sense that it is defined, up to codimension 2, as the scheme resulting from adjoining to $\mathfrak{M}(k,\mu_N)$ roots of explicitly given modular forms. The notation is as in 7.2.

9.1 Definition. *Let* $\overline{\mathfrak{M}}(k,\mu_N)$ *be the minimal compactification of* $\mathfrak{M}(k,\mu_N)$ *constructed in* [Ch, Thm. 4.3]. *It is a projective normal scheme over k obtained by adding finitely many cusps. Its singular locus consists precisely of the complement of the Rapoport locus* $\mathfrak{M}(k,\mu_N)^{\mathrm{R}}$ *defined in 3.5.*

Define

$$\overline{\phi}\colon \overline{\mathfrak{M}}(k,\mu_{pN})^{\mathrm{Kum}} \longrightarrow \overline{\mathfrak{M}}(k,\mu_N)$$

as the normal closure of $\overline{\mathfrak{M}}(k,\mu_N)$ *in* $\mathfrak{M}(k,\mu_{pN})^{\mathrm{Kum}}$ *via the Galois cover ϕ with group $G = G_{1,1}$, defined in 7.2.*

9.2 Lemma. *The following properties hold:*

1. *the morphism $\overline{\phi}$ is finite;*
2. *the scheme* $\overline{\mathfrak{M}}(k,\mu_{pN})^{\mathrm{Kum}}$ *is projective, irreducible and normal;*
3. *the scheme* $\overline{\mathfrak{M}}(k,\mu_{pN})^{\mathrm{Kum}}$ *is endowed with an action of G and $\overline{\phi}$ represents the quotient map;*
4. *the branch locus of $\overline{\phi}$ is a divisor contained in the complement of* $\overline{\mathfrak{M}}(k,\mu_N)^{\mathrm{ord}}$. *We use the convention that the cusps are in the ordinary locus.*

Proof: Since $\overline{\mathfrak{M}}(k,\mu_N)$ is of finite type over k, it is excellent and, hence, universally japanese. By [EGA IV^2, §7.8.3] we conclude that $\overline{\phi}$ is finite. By [EGA II, Cor. 6.1.11] we deduce that $\overline{\phi}$ is projective and, by [EGA II, Prop. 5.5.5 (ii)], that $\overline{\mathfrak{M}}(k,\mu_{pN})^{\mathrm{Kum}}$ is projective. The quotient of $\overline{\mathfrak{M}}(k,\mu_{pN})^{\mathrm{Kum}}$ by G is finite and birational over $\overline{\mathfrak{M}}(k,\mu_N)$. We deduce that it coincides with $\overline{\mathfrak{M}}(k,\mu_N)$. This concludes the proofs of claims (1)-(3). By purity of branch locus, see [SGA 2, X, Thm. 3.4 (i)], the map $\overline{\phi}$ is ramified along a divisor of $\overline{\mathfrak{M}}(k,\mu_N)$. By construction the pre-image of $\mathfrak{M}(k,\mu_N)^{\mathrm{ord}}$ in $\overline{\mathfrak{M}}(k,\mu_{pN})^{\mathrm{Kum}}$ is $\mathfrak{M}(k,\mu_{pN})^{\mathrm{Kum}}$. Hence, $\overline{\phi}$ is étale over $\mathfrak{M}(k,\mu_N)^{\mathrm{ord}}$. Since the cusps are isolated points in the complement of $\mathfrak{M}(k,\mu_N)^{\mathrm{ord}}$, the map $\overline{\phi}$ is unramified also at the cusps. This proves part (4).

9.3 *Local charts of* $\overline{\mathfrak{M}}(k,\mu_{pN})^{\mathrm{Kum}}$. Fix a prime $\mathfrak{Q}$ of O_L over p and an integer $1\leq j\leq f_{\mathfrak{Q}}$. Let

$$\overline{\mathfrak{M}}(k,\mu_N)^{\mathrm{R}}\hookrightarrow\overline{\mathfrak{M}}(k,\mu_N),$$

be the locus where condition (R) holds. The convention is that the cusps satisfy (R). We are going to give an explicit description of $\overline{\phi}$ in a neighborhood of the generic point of the divisor $W_{\mathfrak{Q},j}$ defined by the partial Hasse invariant $h_{\mathfrak{Q},j}$; see 8.18.

Define a scheme

$$\overline{\phi}^{\mathrm{R}}_{\mathfrak{Q},j}\colon\overline{\mathfrak{M}}(k,\mu_{pN})^{\mathrm{Kum,R}}_{\mathfrak{Q},j}\longrightarrow\overline{\mathfrak{M}}(k,\mu_N)^{\mathrm{R}}\setminus\sum_{(\mathfrak{P},i)\neq(\mathfrak{Q},j)}W_{\mathfrak{P},i}$$

over the complement in $\overline{\mathfrak{M}}(k,\mu_N)^{\mathrm{R}}$ of the divisor $\sum_{(\mathfrak{P},i)\neq(\mathfrak{Q},j)}W_{\mathfrak{P},i}$ by adjoining a $p^{f_{\mathfrak{P}}}-1$-th root of the modular forms $h_{\mathfrak{P},i+1}^{p^{f_{\mathfrak{P}}-1}}h_{\mathfrak{P},i+2}^{p^{f_{\mathfrak{P}}-2}}\cdots h_{\mathfrak{P},i}$ for any prime $\mathfrak{P}$ of O_L over p and a fixed integer $1\leq i\leq f_{\mathfrak{P}}$ (we require $i=j$ if $\mathfrak{P}=\mathfrak{Q}$).

1) The identity

$$\left(h_{\mathfrak{P},i+1}^{p^{f_{\mathfrak{P}}-1}}h_{\mathfrak{P},i+2}^{p^{f_{\mathfrak{P}}-2}}\cdots h_{\mathfrak{P},i}\right)^{p}\left(h_{\mathfrak{P},i+2}^{p^{f_{\mathfrak{P}}-1}}h_{\mathfrak{P},i+3}^{p^{f_{\mathfrak{P}}-2}}\cdots h_{\mathfrak{P},i+1}\right)^{-1}=h_{\mathfrak{P},i}^{p^{f_{\mathfrak{P}}}-1}$$

for all $1\leq i\leq f_{\mathfrak{P}}$ implies that the construction is independent of i for $\mathfrak{P}\neq\mathfrak{Q}$.

2) For every prime $\mathfrak{P}$ of O_L over p and any integer $1\leq i\leq f_{\mathfrak{P}}$ we have the following equality of modular forms on $\mathfrak{M}(k,\mu_{pN})^{\mathrm{Kum}}$:

$$\begin{aligned}a(\chi_{\mathfrak{P},i})^{p^{f_{\mathfrak{P}}}-1}&=a(\chi_{\mathfrak{P},i}^{p}\chi_{\mathfrak{P},i+1}^{-1})^{p^{f_{\mathfrak{P}}-1}}a(\chi_{\mathfrak{P},i+1}^{p}\chi_{\mathfrak{P},i+2}^{-1})^{p^{f_{\mathfrak{P}}-2}}\cdots a(\chi_{\mathfrak{P},i-1}^{p}\chi_{\mathfrak{P},i}^{-1})^{p^{0}}\\&=\phi^*(h_{\mathfrak{P},i+1})^{p^{f_{\mathfrak{P}}-1}}\phi^*(h_{\mathfrak{P},i+2})^{p^{f_{\mathfrak{P}}-2}}\cdots\phi^*(h_{\mathfrak{P},i}).\end{aligned}$$

See 7.8 for the notation. In particular, we have a commutative diagram

$$\begin{array}{ccc}\mathfrak{M}(k,\mu_{pN})^{\mathrm{Kum}}&\longrightarrow&\overline{\mathfrak{M}}(k,\mu_{pN})^{\mathrm{Kum,R}}_{\mathfrak{Q},j}\\ \phi\big\downarrow&&\big\downarrow\overline{\phi}^{\mathrm{R}}_{\mathfrak{Q},j}\\ \mathfrak{M}(k,\mu_N)^{\mathrm{ord}}&\longrightarrow&\overline{\mathfrak{M}}(k,\mu_N)^{\mathrm{R}}\setminus\sum_{(\mathfrak{P},i)\neq(\mathfrak{Q},j)}W_{\mathfrak{P},i}.\end{array}$$

By 7.2 the left hand side is a finite étale morphism, Galois under the group G. In particular, its degree is equal to $\prod_{\mathfrak{P}|p}(p^{f_{\mathfrak{P}}}-1)$.

9.4 Proposition. *The scheme* $\overline{\mathfrak{M}}(k,\mu_{pN})^{\mathrm{Kum,R}}_{\mathfrak{Q},j}$ *has the following properties*

1) *it is irreducible and normal;*

2) *the morphism* $\overline{\phi}^{\mathrm{R}}_{\mathfrak{Q},j}$ *is finite, its branch locus is* $(p^{f_{\mathfrak{Q}}}-2)W_{\mathfrak{Q},j}$ *and the inverse image of any irreducible component of* $W_{\mathfrak{Q},j}$ *is irreducible;*

3) *there is a commutative diagram*

$$\begin{array}{ccccc}
\mathfrak{M}(k,\mu_{pN})^{\mathrm{Kum}} & \hookrightarrow & \overline{\mathfrak{M}}(k,\mu_{pN})^{\mathrm{Kum,R}}_{\mathfrak{Q},j} & \hookrightarrow & \overline{\mathfrak{M}}(k,\mu_{pN})^{\mathrm{Kum}} \\
\phi\downarrow & & \overline{\phi}^{R}_{\mathfrak{Q},j}\downarrow & & \downarrow\overline{\phi} \\
\mathfrak{M}(k,\mu_N)^{\mathrm{ord}} & \hookrightarrow & \overline{\mathfrak{M}}(k,\mu_N)^{\mathrm{R}}\backslash \sum_{(\mathfrak{P},i)\neq(\mathfrak{Q},j)} W_{\mathfrak{P},i} & \hookrightarrow & \overline{\mathfrak{M}}(k,\mu_N),
\end{array}$$

where the squares are cartesian and the horizontal arrows are open immersions.

Proof: The finiteness in claim (2) clearly follows from the construction. We prove claims (1)–(2), and claim (3) for the square on the left hand side. Then, the existence of the diagram and the rest of claim (3) are deduced from the definition of $\overline{\mathfrak{M}}(k,\mu_{pN})^{\mathrm{Kum}}$ and the finiteness in (2).

Let x be a closed point of $W_{\mathfrak{Q},j}$ such that $x \notin W_{\mathfrak{P},i}$ for $(\mathfrak{P},i) \neq (\mathfrak{Q},j)$. Let $U_x = \mathrm{Spec}(A)$ be an affine open neighborhood of x in $\overline{\mathfrak{M}}(k,\mu_N)^{\mathrm{R}}\backslash \sum_{(\mathfrak{P},i)\neq(\mathfrak{Q},j)} W_{\mathfrak{P},i}$ over which every invertible sheaf $\mathcal{L}_{\chi_{\mathfrak{P},i}}$ is trivial. Choose an ordering $\mathfrak{Q} = \mathfrak{P}_1 < \cdots < \mathfrak{P}_s$ of the primes of O_L over p. Define

$$B := A\Big[X_1,\ldots,X_s\Big]/\Big(X_t^{p^{f_{\mathfrak{P}_t}}-1} - \big(h_{\mathfrak{P}_t,i+1}\big)^{p^{f_{\mathfrak{P}_t}}} \cdots \big(h_{\mathfrak{P}_t,i}\big)\Big)_{t=1,\ldots,s},$$

with the abuse of notation that, via the chosen trivialization, the elements $h_{\mathfrak{P},i}$ are now considered as elements of A. As remarked in 9.3 the definition does not depend on the choice of $1 \leq i \leq f_{\mathfrak{P}_t}$ for $t > 1$. Then:

i) B is finite and flat over A of degree $\prod_{t=1}^{s}\left(p^{f_{\mathfrak{P}_t}}-1\right)$;

ii) the group $\prod_{t=1}^{s} \mathrm{k}_{\mathfrak{P}_t}^*$ acts A-linearly on B through roots of unity.

Note that $\phi^{-1}\left(U_x \backslash W_{\mathfrak{Q},j}\right) \to U_x \backslash W_{\mathfrak{Q},j}$ is endowed with an action of $G_{1,1} \cong \prod_{t=1}^{s} \mathrm{k}_{\mathfrak{P}}^*$ as remarked in 9.3. By 7.8 the morphism $\phi^{-1}\left(U_x \backslash W_{\mathfrak{Q},j}\right) \to \mathrm{Spec}(B)$ defined in (2) of 9.3 is equivariant with respect to the action of $\prod_{t=1}^{s} \mathrm{k}_{\mathfrak{P}}^*$. Hence, the map

$$\phi^{-1}\left(U_x \backslash W_{\mathfrak{Q},j}\right) \longrightarrow \mathrm{Spec}\left(B\left[h_{\mathfrak{Q},j}^{-1}\right]\right)$$

is an isomorphism. In particular, we conclude that B is a domain. Define

$$B_1 := A\big[X_1\big]/\Big(X_1^{p^{f_{\mathfrak{Q}}}-1} - \big(h_{\mathfrak{Q},i+1}\big)^{p^{f_{\mathfrak{Q}}}} \cdots \big(h_{\mathfrak{Q},i}\big)\Big).$$

We know it is a domain. Let

$$Y = \sum_{d=0}^{p^{f_{\mathfrak{Q}}}-2} Y_d X_t^d \qquad \text{with } Y_d \in \mathrm{Frac}\,(A)$$

be an element in the fraction field of B_1 which is a zero of a monic polynomial $g(X) \in A[X]$. For every prime Q of A, the equation

$$X_1^{p^{f_{\mathfrak{Q}}}-1} - \big(h_{\mathfrak{Q},i+1}\big)^{p^{f_{\mathfrak{Q}}}} \cdots \big(h_{\mathfrak{Q},i}\big)$$

in the local ring A_Q is either Eisenstein, if $h_{\mathfrak{Q},j} \in Q$, or separable otherwise. Hence, $Y_d \in A_Q$ for every d. Since A is normal, we conclude that $Y_d \in A$ for every d. In particular, B_1 is normal. Moreover, the extension $A \subset B_1$ is ramified only along $h_{\mathfrak{Q},j} = 0$ with ramification index $p^{f_{\mathfrak{Q}}} - 1$. The extension $B_1 \subset B$ is étale. Hence, B is normal and $A \subset B$ is ramified only along $h_{\mathfrak{Q},j} = 0$ with ramification index $p^{f_{\mathfrak{Q}}} - 1$. Let Q be a prime ideal of A corresponding to an irreducible component of $W_{\mathfrak{Q},j}$. By 8.17 the $h_{\mathfrak{P},i}$ are local parameters in A. Thus, the extension $B_{1Q}/(X_1) = A_Q/(h_{\mathfrak{Q},i}) \subset B_Q/(X_1) = A_Q/(h_{\mathfrak{Q},i})\Big[X_2,\ldots,X_s\Big]/\Big(X_t^{p^{f_{\mathfrak{P}_t}}-1} - (h_{\mathfrak{P}_t,i+1})^{p^{f_{\mathfrak{P}_t}}} \cdots (h_{\mathfrak{P}_t,i})\Big)_{t=2,\ldots,s}$ is a domain.

9.5 Corollary. *The scheme* $\cup_{\mathfrak{Q},j}\overline{\mathfrak{M}}(k,\mu_{pN})^{\mathrm{Kum,R}}_{\mathfrak{Q},j}$ *is endowed with an action of* G *so that the map* $\overline{\phi}^{\mathrm{R}}$ *is the quotient map. The open subscheme* $\cup_{\mathfrak{Q},j}\overline{\mathfrak{M}}(k,\mu_{pN})^{\mathrm{Kum,R}}_{\mathfrak{Q},j}$ *of* $\overline{\mathfrak{M}}(k,\mu_{pN})^{\mathrm{Kum}}$ *has codimension 2.*

Proof: The first claim is clear. The second follows from 3.6.

9.6 Corollary. *The branch locus of* $\overline{\phi}$ *is exactly the complement of the ordinary locus in* $\overline{\mathfrak{M}}(k,\mu_N)$. *For each prime* $\mathfrak{Q}$ *and each* $1 \le i \le f_{\mathfrak{Q}}$, *the ramification index of* $W_{\mathfrak{Q},i}$ *in* $\overline{\mathfrak{M}}(k,\mu_{pN})^{\mathrm{Kum}}$ *is* $p^{f_{\mathfrak{Q}}} - 1$.

Furthermore, let B^0 *be the local ring of an irreducible component* C *of* $W_{\mathfrak{Q},i}$. *The scheme* $\overline{\phi}^{\mathrm{R}\,-1}(C)$ *is irreducible; let* B *be its local ring. The extension* $B^0 \subset B$ *is Galois with Galois group* $\prod_{\mathfrak{P}} \mathrm{k}^*_{\mathfrak{P}}$ *and factors as* $B^0 \subset B^{\mathrm{et}} \subset B$ *where* $B^0 \subset B^{\mathrm{et}}$ *is étale and Galois with Galois group* $\prod_{\mathfrak{P}\neq\mathfrak{Q}} \mathrm{k}^*_{\mathfrak{P}}$ *(corresponding to the equations* $\{X_t^{p^{f_{\mathfrak{P}_t}}-1} - (h_{\mathfrak{P}_t,i+1})^{p^{f_{\mathfrak{P}_t}}} \cdots (h_{\mathfrak{P}_t,i})\}_{t=2,\ldots,s}$) *and* $B^{\mathrm{et}} \subset B$ *is purely ramified with Galois group* $k^*_{\mathfrak{Q}}$ *(corresponding to the equation* $X_1^{p^{f_{\mathfrak{Q}}}-1} - (h_{\mathfrak{Q},i+1})^{p^{f_{\mathfrak{Q}}}} \cdots (h_{\mathfrak{Q},i})$).

Proof: It follows from 9.4.

9.7 Remark. The locus $\mathfrak{M}(k,\mu_N)^{\mathrm{R}}$ is the locus where the modular forms of level μ_N are defined; see 5.1 and 5.4. This is why we need an explicit description of the map $\overline{\phi}$ over such locus, at least up to codimension 2, as given in the Proposition.

10 Congruences mod p^n and Serre's p-adic modular forms.

There is already a notion of p-adic Hilbert modular forms in the literature [Ka4, §1.9], [Hi,§4]. Although this notion is important and useful, the authors of this paper are not aware of a reference that explains how it stands vis-a-vis a more direct approach. Recall that the theory of p-adic modular forms began when Serre introduced the notion of a p-adic modular form of a given level as a q-expansion which is a p-adic uniform limit of q-expansions of classical modular forms of that same level.

In this setting, Katz's approach of defining p-adic modular forms as certain regular functions on a formal scheme obtained from schemes of the sort $\mathfrak{M}(\mathbf{Z}_p,\mu_{p^nN})$ merged nicely with Serre's approach. See [Ka2, Prop. A1.6].

In the Hilbert modular case the development did not follow the same lines. It seems that the interest was initially in p-adic interpolation of special values of L-functions [DeRi], [Ka4]. Later efforts were mainly devoted to understanding the phenomenon of analytic families of Hilbert modular forms and the connection to completed Hecke algebras; the ensuing theory is now known as Hida's theory.

The authors of this paper are interested in following Serre's original approach. Congruences between modular forms imply congruences between their weights that suggest defining a p-adic modular form as a q-expansion that is a p-adic uniform limit of classical modular forms. Such a limit has a well defined weight in the completion $\widehat{\mathbb{X}}$ of $\mathbb{X}$ with respect to a system of subgroups depending on p.

We prove that a p-adic modular form of weight $\chi \in \widehat{\mathbb{X}}$ defined in this fashion is (almost always) the same thing as a p-adic modular form in Katz's approach, which is an eigenfunction of character χ. We remark that p-adic modular forms à la Katz are certain regular functions on the formal scheme

$$\lim_{\infty \leftarrow m} \left(\lim_{n \to \infty} \mathfrak{M}\big(W_m(k), \mu_{p^n N}\big) \right)$$

(see definition 11.4), and thus correspond to ordinary p-adic modular forms in the case $g = 1$, i. e., to p-adic modular forms of growth condition 1.

One virtue of this isomorphism is that the extension of certain derivation operators $\Theta_{\mathfrak{P},i}$ (see 12.38) to p-adic modular forms is easily proven. This yields an ample supply of examples of p-adic modular forms. The results of this section follow the presentation in Serre, [Se].

10.1 *Notation.* In this section we fix a complete discrete valuation ring R with fraction field F of characteristic 0 and residue field k of characteristic p. Let $\mathfrak{m} = (\pi)$ be the maximal ideal of R. Suppose that R is an O_K-algebra where K is a normal closure of L.

10.2 Definition. *Let $f \in \mathbf{M}\big(F, \mu_N, \chi\big)$ be a $\mathfrak{J}$-polarized modular form over F with N not necessarily prime to p. Let $\big(\mathfrak{A}, \mathfrak{B}, \varepsilon, \mathrm{j}\big)$ be a $\mathfrak{J}$-polarized cusp. Consider the q-expansion $f\big(\mathbf{Tate}(\mathfrak{A}, \mathfrak{B}), \varepsilon, \mathrm{j}\big) = a_0 + \sum_{\nu \in (\mathfrak{A}\mathfrak{B})^+} a_\nu q^\nu$ of f at the given cusp. Define*

$$\mathrm{val}(f) := \sup\big\{n \in \mathbf{Z} \,|\, a_\nu \in \mathfrak{m}^n \; \forall \nu\big\} = \inf\big\{\mathrm{val}_\pi(a_\nu)\big\}.$$

10.3 Proposition. *The notation is as in 10.2. We have,* $\mathrm{val}(f) > -\infty$. *Moreover,*

$$\pi^{-\mathrm{val}(f)} f \in \mathbf{M}\big(R, \mu_N, \chi\big)$$

i. e., is a $\mathfrak{J}$-polarized modular form over R.

Proof: To prove that $\mathrm{val}(f) > -\infty$ note that, with the notation of 6.3, we have

$$f\big(\mathbf{Tate}(\mathfrak{A}, \mathfrak{B}), \varepsilon, \mathrm{j}\big) \in F \otimes_{\mathbf{Z}} \mathbf{Z}\big(\big(\mathfrak{A}, \mathfrak{B}, \sigma_\beta\big)\big).$$

In particular, the valuation of the coefficients of the q-expansion of f is bounded from below. Since the q-expansion of $\pi^{-\mathrm{val}(f)} f$ at the given cusp has integral coefficients, we conclude by 6.10 that $\pi^{-\mathrm{val}(f)} f$ is defined over R.

10.4 Lemma. *The number* $\mathrm{val}(f)$ *is independent of the chosen cusp.*

Proof: By 10.3 we may assume that $\mathrm{val}(f) \geq 0$ at any cusp. By the q-expansion principle explained in 6.10 we have that $f\big(\mathbf{Tate}(\mathfrak{A}, \mathfrak{B}), \varepsilon, \mathrm{j}\big)/\pi^n$ has coefficients in R if and only if f/π^n is in $\mathbf{M}\big(R, \mu_N, \chi\big)$.

10.5 Proposition. *Let $N \geq 4$ be an integer prime to p. Let $f_i \in \mathbf{M}\big(R, \mu_N, \chi_i\big)$, $i = 1, 2$, be a $\mathfrak{J}$-polarized modular form of weight χ_i and level μ_N. Suppose that* *their q-expansions at a $\mathfrak{J}$-polarized unramified cusp $(\mathfrak{A},\mathfrak{B},\varepsilon,\mathrm{j})$ in the sense of 6.6 satisfy*

$$f_1\big(\mathbf{Tate}(\mathfrak{A},\mathfrak{B}),\varepsilon,\mathrm{j}\big)\not\equiv 0 \mod \mathfrak{m},$$

and

$$f_1\big(\mathbf{Tate}(\mathfrak{A},\mathfrak{B}),\varepsilon,\mathrm{j}\big)\equiv f_2\big(\mathbf{Tate}(\mathfrak{A},\mathfrak{B}),\varepsilon,\mathrm{j}\big) \mod \mathfrak{m}^n.$$

Then

$$\chi_1\equiv\chi_2 \mod \mathbb{X}_R(n).$$

See 4.11 for the notation $\mathbb{X}_R(n)$.

Proof: Let $\bar{f}_i$ be the image of f_i in $\mathbf{M}(R/\mathfrak{m}^n,\mu_N,\chi_i)$ for $i=1,2$. Consider the forgetful morphism

$$\psi\colon \mathfrak{M}(R/\mathfrak{m}^n,\mu_{p^nN})\longrightarrow \mathfrak{M}(R/\mathfrak{m}^n,\mu_N).$$

It is a Galois cover with group

$$\mathrm{Aut}_{O_L}\big(\mu_{p^n}\otimes \mathrm{D}_L\big)=\big(O_L/p^nO_L\big)^*.$$

Let

$$r_1=r(\bar{f}_1) \qquad \text{and} \qquad r_2=r(\bar{f}_2)$$

be the associated regular functions on $\mathfrak{M}(R/\mathfrak{m}^n,\mu_{p^nN})$ defined in 7.19. The hypothesis guarantees that $r_1=r_2$. Therefore, if $b\in\big(O_L/p^nO_L\big)^*$ is an element of the Galois group, then

$$\chi_1(b)r_1=[b]\,r_1=[b]\,r_2=\chi_2(b)r_2.$$

Hence, $\big(1-\chi_1\chi_2^{-1}(b)\big)r_1=0$. This implies the claim.

10.6 Corollary. *Let $f_i\in\mathbf{M}(F,\mu_N,\chi_i)$ for $i=1,2$ be two $\mathfrak{J}$-polarized modular forms. Assume their q-expansions at a $\mathfrak{J}$-polarized unramified cusp satisfy*

$$f_1\big(\mathbf{Tate}(\mathfrak{A},\mathfrak{B}),\varepsilon,\mathrm{j}\big)\equiv f_2\big(\mathbf{Tate}(\mathfrak{A},\mathfrak{B}),\varepsilon,\mathrm{j}\big) \mod \mathfrak{m}^n.$$

Then

$$\chi_1\equiv\chi_2 \mod \mathbb{X}_R\Big(n-\min\big\{\mathrm{val}(f_1),\mathrm{val}(f_2)\big\}\Big).$$

Proof: By 10.3 we may assume that f_1 and f_2 are defined over R. Let $m_i:=\mathrm{val}(f_i)$ for $i=1,2$. Without loss of generality we may assume that $m_1\le m_2$. Let $F_i:=\pi^{-m_1}f_i$ for $i=1,2$. By assumption $F_i\in\mathbf{M}(R,\mu_N,\chi_i)$ and $F_1\neq 0$ modulo $\mathfrak{m}$ and $F_1\equiv F_2$ modulo $\mathfrak{m}^{n-m_1}$. We conclude using 10.5.

10.7 Corollary. *Let $f\in\mathbf{M}(F,\mu_N,\chi)$ be a $\mathfrak{J}$-polarized modular form. Consider its q-expansion $f\big(\mathbf{Tate}(\mathfrak{A},\mathfrak{B}),\varepsilon,\mathrm{j}\big)=a_0+\sum_\nu a_\nu q^\nu$ at a $\mathfrak{J}$-polarized unramified cusp. Let*

$$n(\chi):=\min\big\{n\in\mathbf{N}|\,\chi\notin\mathbb{X}_R(n)\big\}.$$

Then

$$\mathrm{val}(a_0)\ge -n(\chi)+\mathrm{val}\big(f-a_0\big).$$

Proof: By 10.3 we may assume that f is defined over R. Consider a_0 as a modular form of weight 1 (the trivial character). By corollary 10.6, we have $\chi\equiv 1$ modulo $\mathbb{X}_R\big(\mathrm{val}(f-a_0)-\mathrm{val}(f)\big)$. Hence, $\mathrm{val}_\pi(a_0)\ge\mathrm{val}(f)\ge -n(\chi)+\mathrm{val}\big(f-a_0\big)$ as wanted.

10.8 Definition. *(Serre modular forms) Suppose that R is π-adically complete. A $\mathfrak{J}$-polarized p-adic Hilbert modular form à la Serre over F of level μ_N (N prime to p) is the equivalence class of a Cauchy sequence $\{f_i \in \mathbf{M}(F, \mu_N, \chi_i)\}_{i\in\mathbf{N}}$ of classical modular forms. 'Cauchy' means Cauchy with respect to* val *i. e., that for any $n \in \mathbf{N}$ there exists $m \in \mathbf{N}$ such that*

$$\mathrm{val}(f_i - f_j) \geq n \quad \text{for all } i, j \geq m.$$

Two Cauchy sequences $\{f_i\}_{i\in\mathbf{N}}$ and $\{g_j\}_{j\in\mathbf{N}}$ such that $\mathrm{val}(f_i - g_i) \to \infty$ for $i \to \infty$ are called equivalent. The next lemma shows that this is a well defined concept.

10.9 Lemma. *Let $(\mathfrak{A}_1, \mathfrak{B}_1, \varepsilon_1, \mathrm{j}_1)$ be a $\mathfrak{J}$-polarized unramified cusp. Let $\{f_i\}_i$ be a Cauchy sequence of $\mathfrak{J}$-polarized Hilbert modular forms with respect to the valuation val_1 associated to $(\mathfrak{A}_1, \mathfrak{B}_1, \varepsilon_1, \mathrm{j}_1)$. If $(\mathfrak{A}_2, \mathfrak{B}_2, \varepsilon_2, \mathrm{j}_2)$ is another unramified $\mathfrak{J}$-polarized cusp with associated valuation val_2, then $\{f_i\}_i$ is Cauchy also with respect to val_2.*

Proof: One reduces to the following assertion. Let $g_i \in \mathbf{M}(R, \mu_N, \chi_i)$, $i = 1, 2$, be modular forms such that $\mathrm{val}_1(g_1 - g_2) \geq n$ then $\mathrm{val}_2(g_1 - g_2) \geq n$.

Viewing g_i as a modular form on $\mathfrak{M}(R/(\pi^n), \mu_{p^n N})^{\mathrm{Kum}}$ we find that $g_1 - g_2$ belongs to the kernel of the q-expansion relative to the cusp $(\mathfrak{A}_1, \mathfrak{B}_1, \varepsilon_1, \mathrm{j}_1)$ if and only if $\mathrm{val}_1(g_1 - g_2) \geq n$. The conclusion follows from Corollary 7.21.

10.10 Definition. *(Weight and q-expansions of p-adic modular forms) Let*

$$f = \{f_i \in \mathbf{M}(F, \mu_N, \chi_i)\}_{i\in\mathbf{N}}$$

be a $\mathfrak{J}$-polarized p-adic Hilbert modular form à la Serre over R of level μ_N. Define the weight $\chi \in \widehat{\mathbb{X}}_R$ of f as

$$\chi := \lim_{i\to\infty} \chi_i \in \widehat{\mathbb{X}}_R.$$

Fix a $\mathfrak{J}$-polarized unramified cusp $(\mathfrak{A}, \mathfrak{B}, \varepsilon, \mathrm{j})$ over F. Define the q-expansion of f at the given cusp by

$$f(\mathbf{Tate}(\mathfrak{A}, \mathfrak{B}), \varepsilon, \mathrm{j}) := \lim_{i\to\infty} f_i(\mathbf{Tate}(\mathfrak{A}, \mathfrak{B}), \varepsilon, \mathrm{j}).$$

Finally, define

$$\mathrm{val}(f) := \sup\{n \in \mathbf{Z} |\, a_\nu \in \mathfrak{m}^n\, \forall \nu\} = \inf\{\mathrm{val}_\pi(a_\nu)\}.$$

10.11 Proposition. *The notation is as in the Definition.*

1) *The weight and the q-expansion at the $\mathfrak{J}$-polarized cusp $(\mathfrak{A}, \mathfrak{B}, \varepsilon, \mathrm{j})$ of a $\mathfrak{J}$-polarized p-adic modular form f à la Serre are well defined i. e., the limits exist and do not depend on the choice of Cauchy sequence of classical modular forms f_i defining it;*
2) *the map*

$$\{p\text{-adic Hilbert modular forms of wt } \chi\} \longrightarrow \{q\text{-expansions at } (\mathfrak{A}, \mathfrak{B}, \varepsilon, \mathrm{j})\},$$

associating to a p-adic Serre modular form f its q-expansion $f(\mathbf{Tate}(\mathfrak{A}, \mathfrak{B}), \varepsilon, \mathrm{j})$, is injective;

3) *the assertions in 10.3-10.7 hold if one replaces* $\mathfrak{I}$*-polarized Hilbert modular forms of level* μ_N *with* $\mathfrak{I}$*-polarized p-adic Hilbert modular forms à la Serre of level* μ_N *and* $\mathbb{X}_R$ *with* $\widehat{\mathbb{X}}_R$.

Proof: Assertions (1) and (2) follow from the results above. The last assertion follows as in [Se, §1].

10.12 Remark. See 11.13 for examples of how Hilbert modular forms of level μ_{Np^n} and trivial nebentypus at p define $\mathfrak{I}$-polarized p-adic Hilbert modular forms. See 18.8 for examples of p-adic, but not classical, $\mathfrak{I}$-polarized Hilbert modular forms arising from Eisenstein series. Other examples are given in 12.27 by applying suitable p-adic theta operators to classical Hilbert modular forms.

10.13 Definition. *We say that a* $\mathfrak{I}$*-polarized p-adic modular form* $\{f_i\}_i$ *of level* μ_N *over* F *is a cusp form if the constant coefficient of its q-expansion at any cusp is* 0.

11 Katz's p-adic Hilbert modular forms.

The notation is as in 10.1.

11.1 Definition. *Let* $m \geq 1$, $n \geq 0$ *and* $N \geq 4$ *be integers. Let* p *be a prime not dividing* N. *Consider the affine schemes*

$$\mathfrak{M}(m,n) = \mathfrak{M}\big(R/\mathfrak{m}^m, \mu_{p^n N}\big).$$

For $n = 0$ *we use the convention* $\mathfrak{M}(m,0) = \mathfrak{M}(R/\mathfrak{m}^m, \mu_N)^{\mathrm{ord}}$. *For* $m' \leq m$ *we have a closed immersion*

$$\mathfrak{M}(m',n) \hookrightarrow \mathfrak{M}(m,n).$$

For $n' \geq n$ *we have a Galois covering*

$$\mathfrak{M}(m,n') \longrightarrow \mathfrak{M}(m,n)$$

with Galois group

$$\Gamma(n',n) = \big(O_L/p^{n'}O_L\big)^*/\big(O_L/p^nO_L\big)^*,$$

where the group $\big(O_L/p^0\,O_L\big)^*$ *is understood as the trivial group. Define*

$$\Gamma_{n'} := \Gamma(n',0) = \mathrm{Aut}_{O_L}(\mu_{p^{n'}} \underset{\mathbf{Z}}{\otimes} \mathrm{D}_L^{-1}) = \big(O_L/p^{n'}\,O_L\big)^*.$$

Define $\mathfrak{M}(\infty,\infty)$ *as the formal scheme*

$$\mathfrak{M}(\infty,\infty) = \lim_{m\to\infty} \mathfrak{M}(m,\infty) = \lim_{m\to\infty}\Big(\lim_{\infty\leftarrow n} \mathfrak{M}(m,n)\Big).$$

The group

$$\Gamma = \lim_{\infty\leftarrow n} \Gamma_n = \lim_{\infty\leftarrow n} \big(O_L/p^n\,O_L\big)^* = \big(\mathbf{Z}_p \underset{\mathbf{Z}}{\otimes} O_L\big)^*$$

acts as Galois automorphisms on $\mathfrak{M}(m,\infty)$ *for every* m *with quotient* $\mathfrak{M}(m,0)$.

Let $n \geq m$. *Let*

$$\mathfrak{M}(m,n)^{\mathrm{Kum}} \longrightarrow \mathfrak{M}(m,0)$$

be the Kummer part of the cover $\mathfrak{M}(m,n) \to \mathfrak{M}(m,0)$; see 7.2. Its Galois group is

$$G_{m,n} \cong \Gamma(n,0)/\big(\cap_\chi \mathrm{Ker}(\chi)\big);$$

the intersection is taken over all the universal characters χ restricted to

$$\chi\colon \big(O_L/p^n O_L\big)^* \longrightarrow \big(R/\mathfrak{m}^m\big)^*.$$

11.2 Lemma. *For every $n \geq m$ the natural map $\mathfrak{M}(m,n)^{\mathrm{Kum}} \to \mathfrak{M}(m,m)^{\mathrm{Kum}}$ is an isomorphism.*

Proof: The space $\mathfrak{M}(m,n)$ is the quotient of $\mathfrak{M}(m,\infty)$ by the action of $1+p^n(\mathbf{Z}_p \otimes O_L)$ while $\mathfrak{M}(m,n)^{\mathrm{Kum}}$ is the quotient of $\mathfrak{M}(m,n)$ by the subgroup $T = \cap \mathrm{Ker}(\chi)$ of Γ_n, the intersection being over all universal characters viewed as homomorphisms $\chi\colon (O_L/p^n O_L)^* \to (R/\mathfrak{m}^m)^*$.

Define T' as $\cap\mathrm{Ker}(\chi)$, where the intersection is taken over all universal (equivalently, basic) characters viewed as homomorphisms $(\mathbf{Z}_p \otimes O_L)^* \to (R/\mathfrak{m}^m)^*$; T' is a subgroup of Γ. Any basic character can be extended to a ring homomorphism $\mathbf{Z}_p \otimes O_L \to R/\mathfrak{m}^m$ and thus is trivial on $p^n(\mathbf{Z}_p \otimes O_L)$ $(n \geq m)$. It follows that $T'/\big(1+p^n(\mathbf{Z}_p \otimes O_L)\big) = T$ and, therefore, $\mathfrak{M}(m,n)^{\mathrm{Kum}}$ is the quotient of $\mathfrak{M}(m,\infty)$ by T' which is independent of n.

11.3 *Weights and characters of* Γ. Let $\chi \in \mathbb{X}_R$ be a character. We may apply χ to Γ by forming the composition

$$\Gamma = (\mathbf{Z}_p \underset{\mathbf{Z}}{\otimes} O_L)^* \hookrightarrow (R \underset{\mathbf{Z}}{\otimes} O_L)^* = \mathcal{G}(R) \xrightarrow{\ \chi\ } R^*.$$

Suppose that $\chi \in \mathbb{X}_R(n)$ (see 4.11), then $\chi(\Gamma) \equiv 1$ modulo $\mathfrak{m}^n$. It follows that every element of $\chi \in \widehat{\mathbb{X}}_R$ defines a well defined homomorphism

$$\Gamma \xrightarrow{\ \chi\ } R^*.$$

11.4 Definition. *(Katz modular forms; c.f. [Ka4, §1.9]) Let χ be a character in $\widehat{\mathbb{X}}_R$. A $\mathfrak{I}$-polarized Katz modular form of weight χ and level μ_N defined over R is a regular function f on $\mathfrak{M}(\infty,\infty)$ such that for every $\alpha \in \Gamma$ we have*

$$\alpha^*(f) = \chi(\alpha)\, f,$$

where $\alpha^*(f) := f \circ \alpha$. Denote the R-module of such functions by

$$\mathbf{M}(R,\mu_N,\chi)^{p-\mathrm{adic}}.$$

The F-module of $\mathfrak{I}$-polarized Katz modular forms of weight χ and level μ_N over F is defined as

$$\mathbf{M}(F,\mu_N,\chi)^{p-\mathrm{adic}} := F \underset{R}{\otimes} \mathbf{M}(R,\mu_N,\chi)^{p-\mathrm{adic}}.$$

11.5 Remark. The notion of $\mathfrak{I}$-polarized Katz p-adic Hilbert modular forms commutes with base change. More precisely, let $R \to R'$ be an extension of O_K-algebras, which are discrete valuation rings of unequal characteristic p and 0. Let F and F' be the associated fraction fields. If $\chi \in \widehat{\mathbb{X}}_R \subset \widehat{\mathbb{X}}_{R'}$, then

$$\mathbf{M}(R',\mu_N,\chi)^{p-\mathrm{adic}} \cong \mathbf{M}(R,\mu_N,\chi)^{p-\mathrm{adic}} \underset{R}{\otimes} R'$$

and

$$\mathbf{M}(F',\mu_N,\chi)^{p-\text{adic}} \cong \mathbf{M}(F,\mu_N,\chi)^{p-\text{adic}} \underset{F}{\otimes} F'.$$

This follows from the fact that the formation of the fine moduli spaces $\mathfrak{M}(m,n)$ commutes with base change.

11.6 Definition. *(q-expansions of Katz modular forms) Define a $\mathfrak{I}$-polarized unramified cusp of $\mathfrak{M}(\infty,\infty)$ to be equivalently*

a) *an unramified cusp $(\mathfrak{A},\mathfrak{B},\varepsilon_{p^\infty N},\mathrm{j}_\varepsilon)$ over R with $\mu_{p^\infty N}$-level structure;*

b) *a compatible system of unramified cusps $(\mathfrak{A},\mathfrak{B},\varepsilon_{p^n N},\mathrm{j}_\varepsilon)$ over $R/\mathfrak{m}^n$.*

Note that we write j_ε in place of $\mathrm{j}_{\varepsilon_{p^\infty}}$ (resp. $\mathrm{j}_{\varepsilon_{p^n}}$). We refer to 6.5 for the latter notation. Given such a cusp, one has a q-expansion map

$$\underset{\chi}{\oplus}\mathbf{M}(F,\mu_N,\chi)^{p-\text{adic}} \longrightarrow F[\![q^\nu]\!]_{\nu\in(\mathfrak{A}\mathfrak{B})^+\cup\{0\}},$$

which is injective; see [Ka4, Thm. 1.10.15]. *With the notation of 6.3, it is defined by evaluating a $\mathfrak{I}$-polarized Katz modular form f defined over R via*

$$\operatorname{Spec}\left(\mathbf{Z}((\mathfrak{A},\mathfrak{B},\sigma_\beta))\underset{\mathbf{Z}}{\otimes} R\right) \longrightarrow \mathfrak{M}(\infty,\infty).$$

We say that a $\mathfrak{I}$-polarized Katz modular form is a cusp form if the constant coefficient of its q-expansion at every cusp is 0.

11.7 *Recall.* If the inequality $m\le n$ holds and $\chi\in\mathbb{X}_R$ there is a canonically defined modular form

$$a(\chi)\in\mathbf{M}\big(R/\mathfrak{m}^m,\mu_{p^n N},\chi\big);$$

see 7.4. Let $(\mathfrak{A},\mathfrak{B},\varepsilon_{p^\infty N},\mathrm{j}_\varepsilon)$ be a $\mathfrak{I}$-polarized unramified cusp of $\mathfrak{M}(m,n)$; see 6.4. Then

i) the modular form $a(\chi)$ descends to a modular form on $\mathfrak{M}\big(R/\mathfrak{m}^m,\mu_{p^n N}\big)^{\mathrm{Kum}}$. See 7.8;

ii) the modular form $a(\chi)$ transforms under Γ_n (equivalently under $G_{m,n}$) according to the character χ^{-1}. See 7.7;

iii) $a(\chi)$ has q-expansion 1 at the cusp $(\mathfrak{A},\mathfrak{B},\varepsilon_{p^\infty N},\mathrm{j}_\varepsilon)$;

iv) for $m'\le m\le n$ and $n'\ge n\ge m$ the modular forms $a(\chi)$ defined on $\mathfrak{M}(m,n)$, on $\mathfrak{M}(m',n)$ and on $\mathfrak{M}(m,n')$ (resp. on $\mathfrak{M}(m,n)^{\mathrm{Kum}}$, on $\mathfrak{M}(m',n)^{\mathrm{Kum}}$ and on $\mathfrak{M}(m,n')^{\mathrm{Kum}}$) agree.

11.8 *The comparison between Serre and Katz modular forms.* Let $g=\{g_i\}_i$ be a $\mathfrak{I}$-polarized Serre p-adic modular form of weight χ, level μ_N and defined over R as in 10.8. We may assume w.l.o.g. that

$$g_{n+1}\equiv g_n \qquad \bmod \mathfrak{m}^n.$$

Let χ_n be the weight of g_n so that $\chi=\lim_n\chi_n$ by 10.10. For every $n\in\mathbf{N}$ define

$$f_n := g_n/a(\chi_n)$$

as a regular function on $\mathfrak{M}(n,n)$ and, hence, on $\mathfrak{M}(n,\infty)$. The q-expansion at a cusp of $\mathfrak{M}(n,n)$ of f_n is the q-expansion of g_n. It follows that $f_{n+1}\equiv f_n \bmod \mathfrak{m}^n$ and, thus, $\{f_n\}_n$ defines a $\mathfrak{I}$-polarized Katz p-adic modular form f of weight χ

and level μ_N over R. Furthermore, the q-expansion of f at a cusp of $\mathfrak{M}(\infty,\infty)$ coincides with that of g. Thus, we obtain a map

$$\{\mathfrak{I}\text{-pol. Serre } p\text{-adic HMF over } F\} \xrightarrow{\ r\ } \{\mathfrak{I}\text{-pol. Katz } p\text{-adic HMF over } F\},$$

which preserves weights. We also conclude that for any $\mathfrak{I}$-polarized unramified cusp $(\mathfrak{A},\mathfrak{B},\varepsilon_{p^\infty N},\mathrm{j}_\varepsilon)$ of $\mathfrak{M}(\infty,\infty)$ the following diagram is commutative

$$\begin{array}{ccc} \{\text{Serre } p\text{-adic HMF over } F\} & \xrightarrow{\ r\ } & \{\text{Katz } p\text{-adic HMF over } F\} \\ q\text{-exp} \searrow & & \swarrow q\text{-exp} \\ & F[\![q^\nu]\!]_{\nu\in(\mathfrak{A}\mathfrak{B})^+\cup\{0\}}. & \end{array}$$

We deduce from 11.6, and the fact that r preserves weights, that the q-expansion map is injective also on the graded ring of $\mathfrak{I}$-polarized Serre p-adic Hilbert modular forms.

11.9 Lemma. *Let $N\geq 4$ be an integer.*

1) *For a suitable integer $n_0>0$ the modular form h^{n_0} (the n_0-th power of the Hasse invariant defined in 7.12) admits a lift to a modular form $\hbar$ over $\mathbf{Z}_p$ of weight $\mathbf{Nm}^{(p-1)n_0}$. We may choose $\hbar$ so that the leading coefficient of its q-expansion at a given cusp is 1.*
2) *For any $\mathfrak{I}$-polarized modular form $f\in\mathbf{M}\big(R/\mathfrak{m}^n,\mu_N,\mathbf{Nm}^s\big)$ there exists a $\mathfrak{I}$-polarized modular form $g_n\in\mathbf{M}\big(R,\mu_N,\mathbf{Nm}^{s'}\big)$ such that*
   2.a) *g_n mod $\mathfrak{m}^n$ and f have the same q-expansion at one (any) cusp;*
   2.b) *$\mathbf{Nm}^s\equiv\mathbf{Nm}^{s'}$ mod $\mathbb{X}_R(n)$.*
3) *For any character $\chi\in\mathbb{X}_R$, any $n\in\mathbf{N}$ and any $\mathfrak{I}$-polarized cusp form $f\in\mathbf{M}\big(R/\mathfrak{m}^n,\mu_N,\chi\big)$ there exists a $\mathfrak{I}$-polarized modular form $g_n\in\mathbf{M}\big(R,\mu_N,\chi'\big)$ such that*
   3.a) *g_n mod $\mathfrak{m}^n$ and f have the same q-expansion at one (any) cusp;*
   3.b) *$\chi\equiv\chi'$ mod $\mathbb{X}_R(n)$.*

Proof: We fix some notation. Let

$$\delta\colon\overline{\mathfrak{M}}(R,\mu_N)_{\sigma_\beta}\longrightarrow\overline{\overline{\mathfrak{M}}}(R,\mu_N)$$

be the morphism from a smooth toroidal compactification to the minimal compactification of $\mathfrak{M}(R,\mu_N)$. See [Ra, §5] and [Ch, §4]. Let

$$\pi\colon\mathrm{A}\longrightarrow\mathfrak{M}(R,\mu_N)$$

be the universal $\mathfrak{I}$-polarized abelian scheme with real multiplication by O_L. By [Ra, §5.4], the abelian scheme A extends to a semiabelian scheme over $\overline{\mathfrak{M}}(R,\mu_N)_{\sigma_\beta}$ such that the sheaf $\pi_*\Omega^1_{\mathrm{A}/\mathfrak{M}(R,\mu_N)}$ extends to a locally free $O_{\overline{\mathfrak{M}}(R,\mu_N)_{\sigma_\beta}}$-module Ω on $\overline{\mathfrak{M}}(R,\mu_N)_{\sigma_\beta}$. The line bundle

$$\wedge^g\pi_*\left(\Omega^1_{\mathrm{A}/\mathfrak{M}(R;\mu_N)}\right)=\mathcal{L}_{\mathbf{Nm}}$$

is the determinant of the Hodge bundle; see 5.4 for the notation. It extends to a line bundle $\wedge^g\pi_*(\Omega)$ on $\overline{\mathfrak{M}}(R,\mu_N)_{\sigma_\beta}$. By [Ch, Thm. 4.3 (IX)] the latter descends to an ample line bundle on $\overline{\overline{\mathfrak{M}}}(R,\mu_N)$, which we denote in the same way. This implies that there exists an integer $n_0\gg 0$ such that $\mathcal{L}_{\mathbf{Nm}^{n_0(p-1)}}$ is very ample. Note that h is a section of $\mathcal{L}_{\mathbf{Nm}}$ over $\mathfrak{M}(k,\mu_N)^{\mathrm{R}}$ which has codimension at least

two 2 in $\overline{\mathfrak{M}}(k, \mu_N)$. By [Ch, Thm. 4.3 (V)], the scheme $\overline{\mathfrak{M}}(k, \mu_N)$ is normal. In particular, h extends to a section over $\overline{\mathfrak{M}}(k, \mu_N)$. Hence, h^{n_0} lifts to a modular form $\hbar$ of $\mathcal{L}_{\mathbf{Nm}^{n_0(p-1)}}$ over $\overline{\mathfrak{M}}(R, \mu_N)$.
Fix a $\mathfrak{I}$-polarized unramified cusp $(\mathfrak{A}, \mathfrak{B}, \varepsilon, \mathrm{j}_\varepsilon)$ of $\mathfrak{M}(R, \mu_N)$. The q-expansion principle implies that the q-expansion of $\hbar$ is congruent to that of h^{n_0} modulo $\mathfrak{m}$. By 7.14, letting

$$a_0 + \sum_{\nu \in (\mathfrak{A}\mathfrak{B})^+} a_\nu q^\nu$$

be the q-expansion of $\hbar$ at the given cusp, we have that $a_0 \equiv 1 \bmod \mathfrak{m}$ and $a_\nu \in \mathfrak{m}$ for all $\nu \neq 0$. In particular, a_0 is a unit. Replacing $\hbar$ by $a_0^{-1}\hbar$ we may assume $\hbar$ to have a leading Fourier coefficient equal to one at the given cusp. This proves (1).
The proof of (2) is analogous: one replaces f by $f' := f\hbar^n$ for $n \gg 0$, argues that f' extends to the minimal compactification and lifts to R provided n is suitably chosen.

Define

$$\delta^o \colon \overline{\mathfrak{M}}(R, \mu_N)^o_{\sigma_\beta} \longrightarrow \overline{\mathfrak{M}}(R, \mu_N)^o$$

as δ restricted to the complement of the divisor of $\hbar$ in $\overline{\mathfrak{M}}(R, \mu_N)$. Let $\chi \in \mathbb{X}_R$. Since $\Omega^1_{A/\mathfrak{M}(R,\mu_N)}$ extends on $\overline{\mathfrak{M}}(R, \mu_N)^o_{\sigma_\beta}$ as an invertible $O_L \otimes_{\mathbf{Z}} O_{\overline{\mathfrak{M}}(R,\mu_N)^o_{\sigma_\beta}}$-module, proceeding as in 5.4 we see that the invertible sheaf $\mathcal{L}_\chi$ on $\mathfrak{M}(R, \mu_N)$ extends to an invertible sheaf $\bar{\mathcal{L}}_\chi$ on $\overline{\mathfrak{M}}(R, \mu_N)^o_{\sigma_\beta}$. Since δ^o is proper, $\delta^o_*(\bar{\mathcal{L}}_\chi)$ is a coherent sheaf. By construction its restriction to $\mathfrak{M}(R, \mu_N)$ coincides with $\mathcal{L}_\chi$. Since $\hbar$ is very ample, $\overline{\mathfrak{M}}(R, \mu_N)^o$ is affine. In particular, for any $n \in \mathbf{N}$ the map

$$\Gamma\left(\overline{\mathfrak{M}}(R, \mu_N)^o, \delta^o_*(\bar{\mathcal{L}}_\chi)\right) \longrightarrow \Gamma\left(\overline{\mathfrak{M}}(R/\mathfrak{m}^n, \mu_N)^o, \delta^o_*(\bar{\mathcal{L}}_\chi)\right)$$

is surjective. Note that

$$\mathbf{M}(R/\mathfrak{m}^n, \mu_N, \chi) = \Gamma\left(\mathfrak{M}(R/\mathfrak{m}^n, \mu_N), \mathcal{L}_\chi \underset{\mathrm{Spec}(R)}{\times} \mathrm{Spec}(R/\mathfrak{m}^n)\right).$$

In particular, if $f \in \mathbf{M}(R/\mathfrak{m}^n, \mu_N, \chi)$ is a cusp form, we can extend f by 0 to a global section f' of $\delta^o_*(\bar{\mathcal{L}}_\chi)$ over $\overline{\mathfrak{M}}(R/\mathfrak{m}^n, \mu_N)^o$. There exists a global section g' of $\delta^o_*(\bar{\mathcal{L}}_\chi)$ over $\overline{\mathfrak{M}}(R, \mu_N)^o$ lifting f'. Hence, there exists $r \gg 0$ such that $g := g'\hbar^r$ extends to a section of $\mathcal{L}_\chi$ over $\mathfrak{M}(R, \mu_N)$ and $\mathbf{Nm}^{(p-1)n_0 r} \equiv 1 \bmod \mathbb{X}_R(n)$. Since

$$\mathbf{M}(R, \mu_N, \chi) = \Gamma\left(\mathfrak{M}(R, \mu_N), \mathcal{L}_\chi\right),$$

this proves (3).

11.10 Lemma. *Let N be an integer. Let $U_1(N)$ denote the elements of O_L^* congruent to 1 modulo N. Let $\chi \in \mathbb{X}_R$ such that we have $\chi(U_1(N)) = 1$. Then χ is a power of* $\mathbf{Nm}$.

Proof: The character χ belongs to $\mathbb{X}$ by 4.2. Let us write the complex embeddings of L as $\sigma_1, \ldots, \sigma_g$. Then we may write $\chi \otimes \mathbf{C} = \sigma_1^{a_1} \ldots \sigma_g^{a_g}$. Replacing χ by χ^2 if necessary, we may assume that the $a_1, \ldots, a_g$ are even. By multiplying χ by a suitable power of $\mathbf{Nm}^2$ we may assume $a_i \geq 0$ for all $1 \leq i \leq g$ and w.l.o.g. $a_1 = 0$. By Dirichlet's units theorem there exists a unit $u \in O_L^*$ such that $\sigma_1(u) > 1$ and $0 < \sigma_i(u) < 1$ for $i = 2, \ldots, g$. Since $U_1(N)$ is of finite index in O_L^*, there exists a power u^n of u such that $u^n \in U_1(N)$. But then $1 = \chi(u^n) = \prod_{i=2}^g \sigma_i^{a_i}(u^n)) \leq 1$. We have equality if and only if $a_i = 0$ for $2 \leq i \leq g$ i. e., χ is a multiple of $\mathbf{Nm}$.

11.11 Theorem. *The notions of a $\mathfrak{I}$-polarized p-adic Hilbert modular form over F in the sense of Serre and in the sense of Katz are the same i. e., the map r is an isomorphism, in the following cases:*

i. *cusp forms;*

ii. *forms of weight $\chi \in \mathbb{X}$;*

iii. *forms of weight $\mathbf{Nm}^z$ with $z \in \mathbf{Z}_p$.*

Moreover, a modular form of non-parallel weight $\chi \in \mathbb{X}$ i. e., whose weight is not of the form $\mathbf{Nm}^s$ for a suitable integer s, is a cusp form.

Proof: The injectivity of r follows from the injectivity of the q-expansion map on Serre p-adic modular forms proven in 10.11. The fact that r preserves the notion of cusp form in the sense of Katz and of Serre follows from 11.8. We are left with the proof of the surjectivity of r. By 10.3 it suffices to prove it for modular forms defined over R.

Let $\{f_n \in O_{\mathfrak{M}(n,\infty)}\}_n$ be a sequence giving a Katz modular form of weight $\chi \in \widehat{\mathbb{X}}$. For every n, the regular function f_n on $\mathfrak{M}(n,\infty)$ is of weight $\chi_n := \chi$ mod $\mathfrak{m}^n$, a character of Γ. In particular, f_n is invariant under $H := \cap_\psi \mathrm{Ker}(\psi)$, where the intersection is taken over all characters $\psi\colon \Gamma \to (R/\mathfrak{m}^n)^*$. Let $\mathfrak{M}(n,\infty) \to \mathfrak{M}(n,\infty)^{\mathrm{Kum}} = \mathfrak{M}(n,n)^{\mathrm{Kum}}$ be the Galois cover with group H. Note that

$$f_n \in O^{\chi_n}_{\mathfrak{M}(n,n)^{\mathrm{Kum}}} \cong \mathcal{L}_{\chi_n}$$

the isomorphism being as $O_L \otimes_{\mathbf{Z}} O_{\mathfrak{M}(n,0)}$-modules. It follows from 7.8 that the modular form $g'_n := a(\chi_n) f_n$ on $\mathfrak{M}(n,n)^{\mathrm{Kum}}$ descends to a modular form on $\mathfrak{M}(n,0)$ of weight χ_n mod. $\mathbb{X}_R(n)$ and with the same q-expansion at any cusp as that of f_n. By multiplying it by a high enough power of the modular form $\hbar$, constructed in 11.9 as a lifting to $\mathbf{Z}_p$ of a power of the Hasse invariant, we may assume that g'_n extends to a modular form defined over $\mathfrak{M}(R/\mathfrak{m}^n, \mu_N)$.

Assume first that f is a cusp form. Then so is each f_n and g'_n. It follows from 11.9 that we may find a modular form g_n over R such that

$$g_n \equiv g'_n \qquad \bmod \mathfrak{m}^n.$$

Hence, the sequence of modular forms g_n of weight $\psi_n \equiv \chi_n$ mod $\mathbb{X}_R(n)$ of modular forms over R converges to a Serre p-adic modular form with the same q-expansion as that of f. It follows from 11.8 that f is the image of the Serre p-adic modular form $\{g_n\}_n$. This proves (i).

Suppose that f is not a cusp form, but has weight $\chi \in \mathbb{X}$. The q-expansion of g'_n lies in the ring

$$(R/\mathfrak{m}^n)[\![q^\nu]\!]^{U_1(N)}_{\nu \in O_L^+},$$

where the action of $U_1(N)$ (the units of O_L congruent to 1 mod N) is given by the "factor of automorphy"

$$q^\nu \mapsto \chi(\epsilon) q^{\epsilon^2 \nu}.$$

Looking at the coefficient q^0, this implies that $\chi(\epsilon) \equiv 1$ mod $\mathfrak{m}^n$ for all n and all $\epsilon \in U_1(N)$ and, therefore, that χ induces the trivial character $\chi\colon U_1(N) \to R^*$. We conclude from 11.10 that χ is of the form $\mathbf{Nm}^r$ for a suitable $r \in \mathbf{Z}$.

In particular, to prove (ii) and (iii) one may assume that $\chi_n = \mathbf{Nm}^{s(n)}$ mod $\mathbb{X}_R(n)$ for a suitable integer $s(n)$. There exists a positive integer t, depending on g'_n, such that the modular form $g'_n \hbar^t$ extends to a modular form on $\mathfrak{M}(R/\mathfrak{m}^n, \mu_N)$. By 11.9, we may take t such that $g_n' \hbar^t$ can be lifted to a modular form g_n of parallel weight, defined over R and whose q-expansion mod $\mathfrak{m}^n$ is the q-expansion of g_n'. The sequence $\{g_n\}_n$ defines a Serre p-adic modular form whose associated Katz p-adic modular form is f by construction.

11.12 Remark. It is interesting to note that although there is no natural morphism $\mathfrak{M}(n,n)^{\mathrm{Kum}} \to \mathfrak{M}(n+1,n+1)^{\mathrm{Kum}}$, the proof above reveals that to give a Katz modular form of level μ_N and weight χ is equivalent to giving a *compatible sequence* of regular functions $\{f_n\}_n$ in $O^{\chi_n}_{\mathfrak{M}(n,n)^{\mathrm{Kum}}}$. The compatibility condition is that $f_{n+1} \equiv f_n \bmod \mathfrak{m}^n$ viewed as regular functions on $\mathfrak{M}(n,\infty)$.

11.13 *Modular forms of level $\Gamma_0(p^n)$ as p-adic modular forms.* Let $\mathbf{M}(R, \mu_{p^n N}, \chi)^{\mathbf{1}}$ be the R-module of $\mathfrak{I}$-polarized modular forms of level $\mu_{p^n N}$, character χ and trivial nebentypus at p i. e., invariant under $\Gamma_n = \mathrm{Aut}\big(\mathrm{D}_L^{-1} \otimes_{\mathbf{Z}} \mu_{p^n}\big)$. Define an F-linear map

$$\tau_F \colon \mathbf{M}(F, \mu_{p^n N}, \chi)^{\mathbf{1}} \longrightarrow \mathbf{M}(F, \mu_N, \chi)^{p-\mathrm{adic}}$$

as follows. For any $m \in \mathbf{N}$, consider the Galois morphism with group Γ_n

$$\psi \colon \mathfrak{M}(R/\mathfrak{m}^m, \mu_{p^n N}) \longrightarrow \mathfrak{M}(R/\mathfrak{m}^m, \mu_N).$$

Let $f \in \mathbf{M}(R, \mu_{p^n N}, \chi)^{\mathbf{1}}$. For any $m \in \mathbf{N}$ the reduction g_m of f defines a modular form on $\mathfrak{M}(R/\mathfrak{m}^m, \mu_{p^n N})$ of weight χ invariant under the action of Γ_n. Hence, g_m descends to a modular form of weight χ on $\mathfrak{M}(m,0) = \mathfrak{M}(R/\mathfrak{m}^m, \mu_N)^{\mathrm{ord}}$ (we freely use the interpretation of modular forms of given weight as sections of line bundles; see 5.4). In particular, for any $s \geq m$ we obtain a regular function

$$\frac{g_m}{a(\chi)} \in \Gamma\left(\mathfrak{M}(m,s), O_{\mathfrak{M}(m,s)}\right);$$

see 11.7 for the notation. This defines a $\mathfrak{I}$-polarized p-adic Hilbert modular form g à la Katz of weight χ. Define

$$\tau_F(f) := g.$$

If f is a $\mathfrak{I}$-polarized modular form of weight χ and level $\mu_{p^n N}$ defined over F, as argued in 10.3, there exists an integer $t \gg 0$ such that $\pi^t f$ is defined over R. Define

$$\tau_F(f) := \pi^{-t} \tau_F(\pi^t f).$$

We now explain the connection to $\mathfrak{I}$-polarized modular forms of level $\mu_N \times \Gamma_0(p^n)$. The notation is as in 5.2. The morphism

$$\mathfrak{M}(\bar{F}, \mu_{p^n N}) \longrightarrow \mathfrak{M}(\bar{F}, \mu_N, \Gamma_0(p^n))$$

is finite, étale and Galois with group $\mathrm{Aut}\big(\mathrm{D}_L^{-1} \otimes_{\mathbf{Z}} \mu_{p^n}\big)$. In particular, we obtain the isomorphism

$$\mathbf{M}(\bar{F}, \mu_N, \Gamma_0(p^n), \chi) \xrightarrow{\sim} \mathbf{M}(\bar{F}, \mu_{p^n N}, \chi)^{\mathbf{1}}.$$

Hence, we get $\bar{F}$-linear maps

$$\mathbf{M}(\bar{F}, \mu_N, \Gamma_0(p^n), \chi) \xrightarrow{\sim} \mathbf{M}(\bar{F}, \mu_{p^n N}, \chi)^{\mathbf{1}} \xrightarrow{\tau_{\bar{F}}} \mathbf{M}(\bar{F}, \mu_N, \chi)^{p-\mathrm{adic}}.$$

By construction we obtain the following important property

$$\tau_{\bar{F}}(f)\Big(\mathbf{Tate}(\mathfrak{A},\mathfrak{B}), \varepsilon_{p^\infty N}, \mathrm{j}_\varepsilon\Big) = f\Big(\mathbf{Tate}(\mathfrak{A},\mathfrak{B}), \varepsilon_{p^n N}, \mathrm{j}_\varepsilon\Big);$$

see 6.6 and 11.6 for the notation.

12 The operators $\Theta_{\mathfrak{P},i}$.

This section is devoted to the construction of certain derivation operators on p-adic Hilbert modular forms and on Hilbert modular forms in characteristic p. For p unramified, these operators were defined by Katz [Ka4, §2.5]. However, our construction is independent; more importantly, in characteristic p the operators defined by Katz are defined only on the ordinary locus, while our operators are defined on the whole moduli space.

Let R be a discrete valuation ring of unequal characteristic p-0 and with maximal ideal $\mathfrak{m}$. Our method is first to define these operators as derivation operators on functions on $\mathfrak{M}\big(R/\mathfrak{m}^n, \mu_{p^n N}\big)$ and then to use the map r of 7.19, which relates modular forms on $\mathfrak{M}\big(R/\mathfrak{m}^n, \mu_N\big)$ to functions on $\mathfrak{M}\big(R/\mathfrak{m}^n, \mu_{p^n N}\big)$, to transport the operators to modular forms on $\mathfrak{M}\big(R/\mathfrak{m}^n, \mu_N\big)^{\mathrm{ord}}$.

In characteristic p, we succeed in defining theta operators on modular forms defined on $\mathfrak{M}\big(R/\mathfrak{m}, \mu_N\big)$. After establishing the existence of these operators and some basic properties we examine how the divisor of a modular form changes under such a derivation operator. This is applied to study how the filtration of a q-expansion changes under these derivation operators.

This section is technically demanding, yet forms the core of §§12-17. To orient the reader we explain its structure.

Sections 12.1-12.6 recall the definition of the Kodaira-Spencer isomorphism in the setting of the schemes $\mathfrak{M}(m,n)$ $(n \geq m)$. The main use we make of this isomorphism is to construct a canonical basis for the holomorphic 1-forms on $\mathfrak{M}(m,n)$ from the modular forms $a(\chi_{\mathfrak{P},i})$ of 7.4.

Sections 12.7-12.11 are devoted to constructing machinery to be used in the following definition of the theta operators. The complications arise from ramification. One of our goals is to have a certain operator Λ – constructed out of theta operators – on modular forms, such that $\Lambda(f)$ is in the image of the operator V. The operator V is essentially raising to a p-power (see Section 13). Hence, we need $\Lambda(f)$ to have q-expansion of the form $\sum_{\nu \in O_L^+} a_{p\nu} q^{p\nu}$. On the other hand, our theta operator associated to a character χ yields a q-expansion of the form $\sum_{\nu \in O_L^+} \chi(\nu) a_\nu q^\nu$ in which "too many" coefficients are zero; we find that in the case of ramification the theta operators need an extra modification for which §§ 12.7-12.9 provide technical background.

Sections 12.12-12.15 provide the definition of the theta operators in the mod p and p-adic settings. In essence, the theta operators are coming from the operation $f \mapsto df$ and the canonical trivialization of the holomorphic one forms on the schemes $\mathfrak{M}(m,n)$. The propagation of this definition to modular forms of level μ_N is carried out later.

Sections 12.16-12.27 are devoted to the calculation of the effect of the theta operators on q-expansions and various corollaries.

Sections 12.28-12.37 are devoted to examining how the poles of a rational function on $\overline{\mathfrak{M}}(m,n)^{\mathrm{Kum}}$, whose poles are supported on the complement of the ordinary locus, change under a theta operator. For this we use the local charts of $\overline{\mathfrak{M}}(m,n)^{\mathrm{Kum}}$ constructed in 9.3. This is used to get well defined theta operators on modular forms and later to determining how the filtration of a modular form changes under a theta operator.

Sections 12.38-12.42 apply the previous results to define and derive the properties of theta operators on modular forms. We remark that on the level of Galois representations, an application of a theta operator corresponds to a twist by a Hecke character. The change of filtration under a theta operator is examined in Section 15; that corresponds to the question of the minimal weight from which a Galois representation arises, up to a twist.

12.1 *Notation.* Let R be a complete discrete valuation ring with maximal ideal $\mathfrak{m}$, residue field k of characteristic p and fraction field F of characteristic 0. Suppose that R is an O_K-algebra where K is a normal closure of L.

12.2 Definition. *Let $n \geq n' \geq m \geq 1$ be integers. Let*

$$\mathfrak{M}(m,n) := \mathfrak{M}(R/\mathfrak{m}^m, \mu_{p^n N}), \qquad \mathfrak{M}(m,n') := \mathfrak{M}(R/\mathfrak{m}^m, \mu_{p^{n'} N}).$$

Let

$$\phi\colon \mathfrak{M}(m,n) \longrightarrow \mathfrak{M}(m,n')$$

be the natural forgetful morphism; it is a Galois cover. We also write

$$\mathfrak{M}(m,0) := \mathfrak{M}(R/\mathfrak{m}^m, \mu_N)^{\mathrm{ord}}.$$

Let

$$d\colon O_{\mathfrak{M}(m,n)} \longrightarrow \Omega^1_{\mathfrak{M}(m,n)/(R/\mathfrak{m}^m)}, \qquad d'\colon O_{\mathfrak{M}(m,n')} \longrightarrow \Omega^1_{\mathfrak{M}(m,n')/(R/\mathfrak{m}^m)}$$

be the derivation from the structure sheaf to the sheaf of differentials of $\mathfrak{M}(m,n)$ (resp. of $\mathfrak{M}(m,n')$) over $R/\mathfrak{m}^m$. Denote by

$$\pi\colon \mathrm{A} \longrightarrow \mathfrak{M}(m,n), \qquad \pi'\colon \mathrm{A}' \longrightarrow \mathfrak{M}(m,n')$$

the universal $\mathfrak{J}$-polarized abelian schemes. Let

$$\omega_{\mathrm{A}/\mathfrak{M}(m,n)} := \pi_*\left(\Omega^1_{\mathrm{A}/\mathfrak{M}(m,n)}\right), \qquad \omega_{\mathrm{A}'/\mathfrak{M}(m,n')} := \pi'_*\left(\Omega^1_{\mathrm{A}'/\mathfrak{M}(m,n')}\right).$$

12.3 Remark. Letting $\mathrm{A}^{\mathrm{U}} \to \mathfrak{M}(m,0)$ be the universal abelian scheme, we have canonical isomorphisms

$$\mathrm{A} \xrightarrow{\sim} \mathrm{A}^{\mathrm{U}} \underset{\mathfrak{M}(m,0)}{\times} \mathfrak{M}(m,n), \qquad \omega_{\mathrm{A}/\mathfrak{M}(m,n)} \xrightarrow{\sim} \psi^*\left(\omega_{\mathrm{A}^{\mathrm{U}}/\mathfrak{M}(m,0)}\right).$$

We deduce that $O_{\mathfrak{M}(m,n)}$, $\Omega^1_{\mathfrak{M}(m,n)/(R/\mathfrak{m}^m)}$ and $\Omega^1_{\mathrm{A}/\mathfrak{M}(m,n)}$ are canonically endowed with an action of Γ_n (see 7.2 for the notation). Note that d is Γ_n-equivariant. It follows from 3.6 that $\omega_{\mathrm{A}/\mathfrak{M}(m,n)}$ is a locally free $O_{\mathfrak{M}(m,n)} \otimes_{\mathbf{Z}} O_L$-module of rank 1 and hence it is endowed with an action of $\left((R/\mathfrak{m}^m) \otimes_{\mathbf{Z}} O_L\right)^*$. Let χ be a universal character. Let $\mathcal{L}_\chi$ the line bundle associated to $\omega_{\mathrm{A}/\mathfrak{M}(m,n)}$ and the character χ as in 5.4. Then $\mathcal{L}_\chi$ is endowed with two actions of $\left(O_L/p^n O_L\right)^*$:

a. the first is induced by the Galois action of Γ_n on $\Omega^1_{\mathrm{A}/\mathfrak{M}(m,n)}$;
b. the second is induced via the natural map

$$\left(O_L/p^n O_L\right)^* \longrightarrow \left((R/\mathfrak{m}^m) \underset{\mathbf{Z}}{\otimes} O_L\right)^*$$

by the action of $\big((R/\mathfrak{m}^m)\otimes_{\mathbf{Z}} O_L\big)^*$. The latter is identified with the automorphism group of $\Omega^1_{\mathrm{A}/\mathfrak{M}(m,n)}$ as $O_{\mathfrak{M}(m,n)}\otimes_{\mathbf{Z}} O_L$-module.

By 7.8 the effect of $\alpha\in\Gamma_n$ on the section $a(\chi)$ is $\chi^{-1}(\alpha)a(\chi)$ via the first action, and $\chi(\alpha)a(\chi)$ via the second action.

12.4 *Notation.* Fix an element $l\in\mathfrak{I}^+$ generating $\mathbf{Z}_p\otimes_{\mathbf{Z}}\mathfrak{I}$ as $\mathbf{Z}_p\otimes_{\mathbf{Z}} O_L$-module as in 7.1. For any $\mathfrak{I}$-polarized abelian scheme with real multiplication by O_L over a $\mathbf{Z}_p$-scheme S such a choice induces a prime to p and O_L-linear polarization.

12.5 Remark. The sequence of differentials

$$0\longrightarrow\pi^*\big(\Omega^1_{\mathfrak{M}(m,n)/(R/\mathfrak{m}^m)}\big)\longrightarrow\Omega^1_{\mathrm{A}/(R/\mathfrak{m}^m)}\longrightarrow\Omega^1_{\mathrm{A}/\mathfrak{M}(m,n)}\longrightarrow 0 \quad (12.5.1)$$

is exact. The exactness on the right and in the middle is obvious; see [EGA IV[4], Cor. 16.4.19]. The exactness on the left follows from the smoothness of A over $\mathfrak{M}(m,n)$; see [EGA IV[4], Prop. 17.2.3]. The sheaves appearing in the sequence above are locally free O_{A}-modules. The functor from the category of sheaves on A, invariant under translation on A, to the category of sheaves on $\mathfrak{M}(m,n)$ (defined by pulling back along the identity section of π) is an equivalence of categories. Hence, there is a canonical isomorphism of O_{A}-modules

$$\pi^*\big(\Omega^1_{\mathfrak{M}(m,n)/(R/\mathfrak{m}^m)}\big)\xrightarrow{\sim}\Omega^1_{\mathfrak{M}(m,n)/(R/\mathfrak{m}^m)}\otimes_{O_{\mathfrak{M}(m,n)}} O_{\mathrm{A}}.$$

Applying the functor π_* to the sequence (12.5.1), we obtain an exact sequence of $O_{\mathfrak{M}(m,n)}$-modules

$$0\longrightarrow\Omega^1_{\mathfrak{M}(m,n)/(R/\mathfrak{m}^m)}\longrightarrow\pi_*\Big(\Omega^1_{\mathrm{A}/(R/\mathfrak{m}^m)}\Big)$$
$$\longrightarrow\omega_{\mathrm{A}/\mathfrak{M}(m,n)}\longrightarrow R^1\pi_*\Big(\pi^*\big(\Omega^1_{\mathfrak{M}(m,n)/(R/\mathfrak{m}^m)}\big)\Big).$$

We have canonical isomorphisms

$$R^1\pi_*\Big(\pi^*\big(\Omega^1_{\mathfrak{M}(m,n)/(R/\mathfrak{m}^m)}\big)\Big)\xrightarrow{\sim}\Omega^1_{\mathfrak{M}(m,n)/(R/\mathfrak{m}^m)}\otimes_{O_{\mathfrak{M}(m,n)}} R^1\pi_*(O_{\mathrm{A}})$$
$$\xrightarrow{\sim}\Omega^1_{\mathfrak{M}(m,n)/(R/\mathfrak{m}^m)}\otimes_{O_{\mathfrak{M}(m,n)}}\mathrm{Hom}_{O_{\mathfrak{M}(m,n)}}\big(\omega_{\mathrm{A}^\vee/\mathfrak{M}(m,n)},O_{\mathfrak{M}(m,n)}\big),$$

where $\mathrm{A}^\vee$ is the dual abelian scheme of A. Call the object on the right hand side N. By 12.4 we get a prime-to-p and O_L-linear polarization $A\to A^\vee$. This induces an isomorphism

$$\omega_{\mathrm{A}/\mathfrak{M}(m,n)}\otimes_{O_{\mathfrak{M}(m,n)}\otimes O_L}\omega_{\mathrm{A}/\mathfrak{M}(m,n)}\cong\omega_{\mathrm{A}/\mathfrak{M}(m,n)}\otimes_{O_{\mathfrak{M}(m,n)}\otimes O_L}\omega_{\mathrm{A}^\vee/\mathfrak{M}(m,n)}$$

and hence a morphism

$$\omega_{\mathrm{A}/\mathfrak{M}(m,n)}^{\otimes^2_{O_L}}\longrightarrow N\otimes_{O_{\mathfrak{M}(m,n)}\otimes O_L}\omega_{\mathrm{A}/\mathfrak{M}(m,n)}\longrightarrow\Omega^1_{\mathfrak{M}(m,n)/(R/\mathfrak{m}^m)}.$$

Here $\otimes^2_{O_L}$ means tensor product as $O_{\mathfrak{M}(m,n)}\otimes O_L$-modules. A similar remark holds for A' in place of A and $\mathfrak{M}(m,n')$ in place of $\mathfrak{M}(m,n)$.

12.6 Proposition. *The maps described above induce canonical isomorphisms, called Kodaira-Spencer isomorphisms,*

$$\mathrm{KS}'\colon \Omega^1_{\mathfrak{M}(m,n')/(R/\mathfrak{m}^m)} \longrightarrow \omega^{\otimes^2_{O_L}}_{\mathrm{A}'/\mathfrak{M}(m,n')}$$

and

$$\mathrm{KS}\colon \Omega^1_{\mathfrak{M}(m,n)/(R/\mathfrak{m}^m)} \longrightarrow \omega^{\otimes^2_{O_L}}_{\mathrm{A}/\mathfrak{M}(m,n)}.$$

In particular, $\Omega^1_{\mathfrak{M}(m,n')/(R/\mathfrak{m}^m)}$ (resp. $\Omega^1_{\mathfrak{M}(m,n)/(R/\mathfrak{m}^m)}$) is endowed with the structure of a free $O_{\mathfrak{M}(m,n')} \otimes_{\mathbf{Z}} O_L$-module (resp. of $O_{\mathfrak{M}(m,n)} \otimes_{\mathbf{Z}} O_L$-module) of rank 1. Moreover, the following diagram is commutative:

$$\begin{array}{ccc}
\Omega^1_{\mathfrak{M}(m,n)/(R/\mathfrak{m}^m)} & \xrightarrow{\mathrm{KS}} & \omega^{\otimes^2_{O_L}}_{\mathrm{A}/\mathfrak{M}(m,n)} \\
\downarrow \wr & & \downarrow \wr \\
\phi^*\left(\Omega^1_{\mathfrak{M}(m,n')/(R/\mathfrak{m}^m)}\right) & \xrightarrow{\phi^*(\mathrm{KS}')} & \omega^{\otimes^2_{O_L}}_{\mathrm{A}'/\mathfrak{M}(m,n')}.
\end{array}$$

In particular, the map KS is Γ_n-equivariantly.

Proof: The fact that KS and KS$'$ are isomorphisms is well known. See [Ka4, §1.0.21]. The rest is clear.

12.7 *Recall.* The notation is as in 7.4. Define

$$\omega^{\mathrm{can}} \otimes \omega^{\mathrm{can}} \in \Gamma\left(\mathfrak{M}(m,n), \omega^{\otimes^2_{O_L}}_{\mathrm{A}/\mathfrak{M}(m,n)}\right)$$

$$(\text{resp. } \mathrm{KS}^{-1}\left(\omega^{\mathrm{can}} \otimes \omega^{\mathrm{can}}\right) \in \Gamma\left(\mathfrak{M}(m,n), \Omega^1_{\mathfrak{M}(m,n)/(R/\mathfrak{m}^m)}\right))$$

the canonical generators of $\omega^{\otimes^2_{O_L}}_{\mathrm{A}/\mathfrak{M}(m,n)}$ (resp. $\Omega^1_{\mathfrak{M}(m,n)/(R/\mathfrak{m}^m)}$) as $O_{\mathfrak{M}(m,n)} \otimes_{\mathbf{Z}} O_L$-module.

For each prime $\mathfrak{P}$ of O_L dividing p take $\pi_{\mathfrak{P}} \in O_L$ to be a generator of the ideal $\mathfrak{P} O_{L,\mathfrak{P}}$ as in 2.1. Let $\mathfrak{p} := \mathfrak{m} \cap O_K$. For each integer $1 \leq i \leq f_{\mathfrak{P}}$, let

$$\sigma_{\mathfrak{P},i}\colon R \underset{\mathbf{Z}}{\otimes} O_L \longrightarrow R \qquad \text{and} \qquad e_{\mathfrak{P},i} \in R \underset{\mathbf{Z}}{\otimes} O_L$$

be as in 2.1. Recall that R is assumed to be an $\mathfrak{m}$-adically complete O_K-algebra.

For every $\mathfrak{j} \in \mathbf{N}$ we denote by

$$\left[(R/\mathfrak{m}^m) \underset{\mathbf{Z}}{\otimes} \mathfrak{P}^{\mathfrak{j}}\right]$$

the image of $(R/\mathfrak{m}^m) \otimes_{\mathbf{Z}} \mathfrak{P}^{\mathfrak{j}}$ in $(R/\mathfrak{m}^m) \otimes_{\mathbf{Z}} O_L$.

12.8 Lemma. *Let $m \geq 1$. Let $\mathfrak{P}$ be a prime of O_L over p and let $0 \leq \mathfrak{j} \leq e_{\mathfrak{P}} - 1$. There exists a maximal non-negative integer $1 \leq t^{[\mathfrak{j}]}_{\mathfrak{P}}(m) \leq m$ satisfying the following. For each $1 \leq i \leq f_{\mathfrak{P}}$ there exists a unique morphism*

$$\tilde{\sigma}^{[\mathfrak{j}]}_{\mathfrak{P},i}\colon \left[(R/\mathfrak{m}^m) \underset{\mathbf{Z}}{\otimes} \mathfrak{P}^{\mathfrak{j}}\right] \longrightarrow (R/\mathfrak{m}^{t^{[\mathfrak{j}]}_{\mathfrak{P}}(m)})$$

of $(R/\mathfrak{m}^m)\otimes_{\mathbf{Z}} O_L$-modules such that the diagram

$$\begin{array}{ccc} (R/\mathfrak{m}^m)\otimes_{\mathbf{Z}} O_L & \xrightarrow{\cdot(1\otimes\pi_{\mathfrak{P}}^{\mathrm{j}})} & \left[(R/\mathfrak{m}^m)\otimes_{\mathbf{Z}} \mathfrak{P}^{\mathrm{j}}\right] \\ \sigma_{\mathfrak{P},i}\downarrow & \swarrow \tilde{\sigma}_{\mathfrak{P},i}^{[\mathrm{j}]} & \\ R/\mathfrak{m}^{t_{\mathfrak{P}}^{[\mathrm{j}]}(m)} & & \end{array}$$

commutes. Moreover, the sequence $\{t_{\mathfrak{P}}^{[\mathrm{j}]}(m)\}_{m\in\mathbf{N}}$ is non-decreasing and

$$\lim_{m\to\infty} t_{\mathfrak{P}}^{[\mathrm{j}]}(m)=\infty.$$

Proof: Fix $0\le \mathrm{j}\le e_{\mathfrak{P}}-1$. The morphism $\tilde{\sigma}_{\mathfrak{P},i}^{[\mathrm{j}]}$, whose existence is claimed in the lemma exists if and only if $\mathrm{Ker}\big(\cdot(1\otimes\pi_{\mathfrak{P}}^{\mathrm{j}})\big)\subset\mathrm{Ker}\big(\sigma_{\mathfrak{P},i}\big)$. Consider first the case $m=1$. Then $R/\mathfrak{m}^1=k$. Moreover, $a\in\mathrm{Ker}\big(\cdot(1\otimes\pi_{\mathfrak{P}}^{\mathrm{j}})\big)$ if and only if $a=(1\otimes\pi_{\mathfrak{P}}^{e_{\mathfrak{P}}-\mathrm{j}})\cdot b$ for some $b\in k\otimes_{\mathbf{Z}} O_L$. Since $e_{\mathfrak{P}}-\mathrm{j}>0$, it follows that $\sigma_{\mathfrak{P},i}(a)=0$. This proves that $t_{\mathfrak{P}}^{[\mathrm{j}]}(1)=1$ and that $t_{\mathfrak{P}}^{[\mathrm{j}]}(m)\ge t_{\mathfrak{P}}^{[\mathrm{j}]}(1)=1$ for any m. Clearly $t_{\mathfrak{P}}^{[\mathrm{j}]}(m+1)\ge t_{\mathfrak{P}}^{[\mathrm{j}]}(m)$. Since $\cdot(1\otimes\pi_{\mathfrak{P}}^{\mathrm{j}})\colon R\otimes_{\mathbf{Z}} O_L\to R\otimes_{\mathbf{Z}} O_L$ is injective and R is $\mathfrak{m}$-adically complete, it is clear that

$$\lim_{\infty\leftarrow m}\mathrm{Ker}\Big(\cdot(1\otimes\pi_{\mathfrak{P}}^{\mathrm{j}})|_{(R/\mathfrak{m}^m)\otimes_{\mathbf{Z}} O_L}\Big)=0.$$

Hence, $\lim_{m\to\infty} t_{\mathfrak{P}}^{[\mathrm{j}]}(m)=\infty$ as claimed.

12.9 *Example.* As far as the theory of Hilbert modular forms modulo p is concerned, only the case $m=1$ is relevant. The general case is used to construct theta operators in the p-adic setting.

If $m=1$, then $t_{\mathfrak{P}}^{[\mathrm{j}]}(1)=1$ and for every $v\in(R/\mathfrak{m})\otimes O_L$ we have $\tilde{\sigma}_{\mathfrak{P},i}^{[\mathrm{j}]}(\pi_{\mathfrak{P}}^{\mathrm{j}}v)=\sigma_{\mathfrak{P},i}(v)$.

12.10 Definition. *Let $\mathfrak{P}$ be a prime of O_L over p and let $1\le i\le f_{\mathfrak{P}}$. Define*

$$\sigma_{\mathfrak{P},i}\colon \omega_{\mathrm{A}/\mathfrak{M}(m,n)}^{\otimes^2_{O_L}}\longrightarrow \mathcal{L}_{\chi^2_{\mathfrak{P},i}}$$

to be the unique morphism of $O_{\mathfrak{M}(m,n)}$-modules such that:

1) *$\sigma_{\mathfrak{P},i}(\omega^{\mathrm{can}}\otimes\omega^{\mathrm{can}})=a\big(\chi^2_{\mathfrak{P},i}\big)$. See 7.8 for the definition of $a\big(\chi_{\mathfrak{P},i}\big)$;*
2) *for any $\alpha\in O_L$ and any local section ω of $\omega_{\mathrm{A}/\mathfrak{M}(m,n)}^{\otimes^2_{O_L}}$ we have*

$$\sigma_{\mathfrak{P},i}(\alpha\cdot\omega)=\sigma_{\mathfrak{P},i}(\alpha)\sigma_{\mathfrak{P},i}(\omega).$$

Let $0\le\mathrm{j}\le e_{\mathfrak{P}}-1$ be an integer. Let $t_{\mathfrak{P}}^{[\mathrm{j}]}(m)$ be as in 12.8. Let

$$\mathcal{L}_{\chi^2_{\mathfrak{P},i}}\longrightarrow\mathfrak{M}\big(t_{\mathfrak{P}}^{[\mathrm{j}]}(m),n\big)$$

be the line bundle defined in 5.4. Define

$$\tilde{\sigma}_{\mathfrak{P},i}^{[\mathrm{j}]}\colon\left[\omega_{\mathrm{A}/\mathfrak{M}(m,n)}^{\otimes^2_{O_L}}\underset{O_L}{\otimes}\mathfrak{P}^{\mathrm{j}}\right]\longrightarrow\mathcal{L}_{\chi^2_{\mathfrak{P},i}}$$

as the unique morphism of $O_L \otimes_{\mathbf{Z}} O_{\mathfrak{M}(m,n)}$*-modules such that the following diagram commutes*

$$\begin{array}{ccc} \omega_{A/\mathfrak{M}(m,n)}^{\otimes^2_{O_L}} & \xrightarrow{\cdot(1\otimes\pi_{\mathfrak{P}}^{\mathbf{j}})} & \left[\omega_{A/\mathfrak{M}(m,n)}^{\otimes^2_{O_L}} \otimes_{O_L} \mathfrak{P}^{\mathbf{j}}\right] \\ \sigma_{\mathfrak{P},i}\downarrow & \swarrow \tilde{\sigma}_{\mathfrak{P},i}^{[\mathbf{j}]} & \\ \mathcal{L}_{\chi^2_{\mathfrak{P},i}} & & \end{array}$$

Its existence is guaranteed by 12.8.

12.11 Lemma. *The morphisms* $\tilde{\sigma}_{\mathfrak{P},i}^{[\mathbf{j}]}$ *satisfy the following properties*

1) *they are compatible for different* m *and* n *i. e., if* $n' \geq n$ *and* $m' \geq m$ *are integers such that* $n' \geq m'$ *and* $n \geq m$*, then*
   1.a) *the morphism* $\tilde{\sigma}_{\mathfrak{P},i}^{[\mathbf{j}]}$*, defined on* $\left[\omega_{A/\mathfrak{M}(m',n')}^{\otimes^2_{O_L}} \otimes_{O_L} \mathfrak{P}^{\mathbf{j}}\right]$*, restricts mod* $\mathfrak{m}^m$ *to the morphism* $\tilde{\sigma}_{\mathfrak{P},i}^{[\mathbf{j}]}$*, defined on* $\left[\omega_{A/\mathfrak{M}(m,n')}^{\otimes^2_{O_L}} \otimes_{O_L} \mathfrak{P}^{\mathbf{j}}\right]$*;*
   1.b) *the morphism* $\tilde{\sigma}_{\mathfrak{P},i}^{[\mathbf{j}]}$*, defined on* $\left[\omega_{A/\mathfrak{M}(m,n')}^{\otimes^2_{O_L}} \otimes_{O_L} \mathfrak{P}^{\mathbf{j}}\right]$*, coincides with the pull-back of the morphism* $\tilde{\sigma}_{\mathfrak{P},i}^{[\mathbf{j}]}$ *defined on* $\left[\omega_{A/\mathfrak{M}(m,n)}^{\otimes^2_{O_L}} \otimes_{O_L} \mathfrak{P}^{\mathbf{j}}\right]$.
2) *Let* Γ_n *be the Galois group of* $\mathfrak{M}(m,n) \to \mathfrak{M}(m,0)$*. For any* $\alpha \in \Gamma_n$*, we have*

$$\alpha^*\left(\tilde{\sigma}_{\mathfrak{P},i}^{[\mathbf{j}]}\right) = \tilde{\sigma}_{\mathfrak{P},i}^{[\mathbf{j}]}.$$

Proof: By the definition of $\tilde{\sigma}_{\mathfrak{P},i}^{[\mathbf{j}]}$, it suffices to prove the lemma for $\mathbf{j} = 0$ i. e., for $\sigma_{\mathfrak{P},i}$. Both $\omega_{A/\mathfrak{M}(m,n)}^{\otimes^2_{O_L}}$ and $\mathcal{L}_{\chi^2_{\mathfrak{P},i}}$ are defined over $\mathfrak{M}(m,0)$ and are compatible for different m's. In particular, they are endowed with a canonical action of Γ_n proving that statement (2) makes sense. By 5.4 the line bundle $\mathcal{L}_{\chi^2_{\mathfrak{P},i}}$ over $\mathfrak{M}(m,0)$ is defined by push-out of $\omega_{A/\mathfrak{M}(m,0)}^{\otimes^2_{O_L}}$ by the map $\sigma_{\mathfrak{P},i}$ of 12.8. By 7.8 this defines the map $\tilde{\sigma}_{\mathfrak{P},i}^{[\mathbf{j}]}$ over $\mathfrak{M}(m,0)$. The conclusions follow.

12.12 Definition. *For each prime* $\mathfrak{P}$ *of* O_L *dividing* p *and each integer* $1 \leq i \leq f_{\mathfrak{P}}$*, define the* $(R/\mathfrak{m}^m)$*-derivation*

$$\Theta_{\mathfrak{P},i}\colon O_{\mathfrak{M}(m,n)} \longrightarrow O_{\mathfrak{M}(m,n)}$$

by the formula

$$f \longmapsto \sigma_{\mathfrak{P},i}\Big(\mathrm{KS}(df)\Big)\cdot a\big(\chi^2_{\mathfrak{P},i}\big)^{-1}.$$

See 12.10 for the definition of $\sigma_{\mathfrak{P},i}$*. For* $0 \leq \mathbf{j} \leq e_{\mathfrak{P}} - 1$ *define the subsheaf of* $O_{\mathfrak{M}(m,n)}$

$$O_{\mathfrak{M}(m,n)}^{\mathfrak{P},[\mathbf{j}]}(U) := \left\{f \in \Gamma(U, O_{\mathfrak{M}(m,n)}) | \, (df) \equiv 0 \bmod \mathfrak{P}^{\mathbf{j}}\right\}.$$

Define the $(R/\mathfrak{m}^m)$*-derivation*

$$\Theta_{\mathfrak{P},i}^{[\mathbf{j}]}\colon O_{\mathfrak{M}(m,n)}^{\mathfrak{P},[\mathbf{j}]} \longrightarrow O_{\mathfrak{M}(t_{\mathfrak{P}}^{[\mathbf{j}]}(m),n)}$$

(both considered as sheaves of rings on $\mathfrak{M}(m,n)$) by the formula

$$f \longmapsto \tilde{\sigma}^{[\mathbf{j}]}_{\mathfrak{P},i}\Big(\mathrm{KS}(df)\Big)\cdot a(\chi^2_{\mathfrak{P},i})^{-1}.$$

See 12.10 for the definition of $\tilde{\sigma}^{[\mathbf{j}]}_{\mathfrak{P},i}$ and 12.8 for the definition of $t^{[\mathbf{j}]}_{\mathfrak{P}}(m)$. Note that

$$O^{\mathfrak{P},[0]}_{\mathfrak{M}(m,n)} = O_{\mathfrak{M}(m,n)}$$

and $\Theta^{[0]}_{\mathfrak{P},i}$ coincides with $\Theta_{\mathfrak{P},i}$.

12.13 Remark. The function $\Theta_{\mathfrak{P},i}(f)$ is thus obtained by taking df, viewing it as a section of the $O_L\otimes O_{\mathfrak{M}(m,n)}$-module $\omega^{\otimes^2_{O_L}}_{A/\mathfrak{M}(m,n)}$, projecting it to the $\chi^2_{\mathfrak{P},i}$ component to get a section of $\mathcal{L}_{\chi^2_{\mathfrak{P},i}}$, and then dividing by the non-vanishing section $a(\chi^2_{\mathfrak{P},i})$ to obtain a regular function on $\mathfrak{M}(m,n)$.

Note that, for example, if $m=1$, $\mathbf{j}>0$ and $f\in O^{\mathfrak{P},[\mathbf{j}]}_{\mathfrak{M}(m,n)}$ then $\Theta_{\mathfrak{P},i}(f)=0$, while $\Theta^{[\mathbf{j}]}_{\mathfrak{P},i}(f)$ (morally obtained from $\Theta_{\mathfrak{P},i}(f)$ by dividing by $\pi^{\mathbf{j}}_{\mathfrak{P}}$) is typically not zero. A similar phenomenon would happen with modular forms. This motivates the introduction of the operators $\Theta^{[\mathbf{j}]}_{\mathfrak{P},i}$.

12.14 Proposition. *Let f be a regular function on $\mathfrak{M}(m,n)$. Suppose that f is an eigenfunction for the action of Γ_n of weight $\psi\colon\Gamma_n\to(R/\mathfrak{m}^m)^*$ i. e., $\alpha^*(f)=\psi(\alpha)f$. Then, $\Theta^{[\mathbf{j}]}_{\mathfrak{P},i}(f)$, whenever defined, is an eigenfunction for Γ_n of weight $\psi\cdot\chi^2_{\mathfrak{P},i}$.*

Proof: For any $\alpha\in\Gamma_n$ denote by α^* the induced action on functions, differentials, etc. Using 12.6 and 7.8, we obtain:

$$\begin{aligned}\psi(\alpha)\mathrm{KS}(df) &= \mathrm{KS}(\psi(\alpha)df) = \mathrm{KS}\left(d(\psi(\alpha)f)\right)\\ &= \mathrm{KS}\left(d(\alpha^*f)\right) = \mathrm{KS}\left(\alpha^*(df)\right)\\ &= \alpha^*\left(\mathrm{KS}(df)\right).\end{aligned}$$

Applying $\tilde{\sigma}^{[\mathbf{j}]}_{\mathfrak{P},i}$ to the first and last terms of these inequalities and using 12.11, we have

$$\begin{aligned}\psi(\alpha)\tilde{\sigma}^{[\mathbf{j}]}_{\mathfrak{P},i}\mathrm{KS}(df) &= \tilde{\sigma}^{[\mathbf{j}]}_{\mathfrak{P},i}(\psi(\alpha)\mathrm{KS}(df))\\ &= \tilde{\sigma}^{[\mathbf{j}]}_{\mathfrak{P},i}\Big(\alpha^*\left(\mathrm{KS}(df)\right)\Big)\\ &= \alpha^*\left(\tilde{\sigma}^{[\mathbf{j}]}_{\mathfrak{P},i}(\mathrm{KS}(df))\right).\end{aligned}$$

Since by 7.8 we have that

$$\alpha^*\left(a(\chi^2_{\mathfrak{P},i})\right) = \chi^{-2}_{\mathfrak{P},i}(\alpha)\cdot a(\chi^2_{\mathfrak{P},i}),$$

the conclusion follows.

12.15 *Theta operators on Katz p-adic modular forms.* The notation is as in 11.4. For every prime ideal $\mathfrak{P}$ of O_L over p and any $1\le i\le f_{\mathfrak{P}}$ one can define the R-linear operator

$$\Theta_{\mathfrak{P},i}\colon \mathbf{M}(R,\mu_N,\chi)^{p-\mathrm{adic}} \longrightarrow \mathbf{M}(R,\mu_N,\chi\chi^2_{\mathfrak{P},i})^{p-\mathrm{adic}}$$

as follows. Let $\{f_n\}_n$ be a compatible sequence of functions as in 11.12 defining a $\mathfrak{J}$-polarized (Katz) p-adic Hilbert modular form f of weight χ. Let $\chi_n = \chi \mod \mathbb{X}_R(n)$ be the weight of f_n. Define

$$\Theta_{\mathfrak{P},i}(f_n) \in \Gamma\big(\mathfrak{M}(n,n), O_{\mathfrak{M}(n,n)}\big)$$

as in 12.12. By the compatibility of the Kodaira-Spencer morphisms proven in 12.6, by 11.7 and 12.11, it follows that $\{\Theta_{\mathfrak{P},i}(f_n)\}_n$ is a compatible sequence defining a p-adic modular form $\Theta_{\mathfrak{P},i}(f)$ of weight $\chi \cdot \chi^2_{\mathfrak{P},i}$.

For any $0 \leq \mathfrak{j} \leq e_{\mathfrak{P}} - 1$ let

$$\mathbf{M}(R,\mu_N,\chi)^{p-\mathrm{adic},\mathfrak{P},[\mathfrak{j}]} := \lim_{\infty \leftarrow m} \lim_{n \to \infty} \Gamma\left(\mathfrak{M}(m,n), O^{\mathfrak{P},[\mathfrak{j}],\chi}_{\mathfrak{M}(m,n)}\right).$$

An element of this space can be described as a compatible sequence of functions $\{f_n\}_n$, $f_n \in \Gamma(\mathfrak{M}(n,n), O^{\mathfrak{P},[\mathfrak{j}]}_{\mathfrak{M}(n,n)})$, each f_n is a χ-eigenfunction for Γ_n; cf. 11.12. One defines the R-linear operator

$$\Theta^{[\mathfrak{j}]}_{\mathfrak{P},i} \colon \mathbf{M}(R,\mu_N,\chi)^{p-\mathrm{adic},\mathfrak{P},[\mathfrak{j}]} \longrightarrow \mathbf{M}(R,\mu_N,\chi\chi^2_{\mathfrak{P},i})^{p-\mathrm{adic},\mathfrak{P},[\mathfrak{j}]}$$

as follows. With the notation of 12.8, we have a compatible sequence

$$\Theta^{[\mathfrak{j}]}_{\mathfrak{P},i}(f_n) \in \Gamma\left(\mathfrak{M}(t^{[\mathfrak{j}]}_{\mathfrak{P}}(n),n), O_{\mathfrak{M}(t^{[\mathfrak{j}]}_{\mathfrak{P}}(n),n)}\right).$$

Since $t^{[\mathfrak{j}]}_{\mathfrak{P}}(n) \to \infty$ this defines a p-adic modular form of weight $\chi\chi^2_{\mathfrak{P},i}$.

12.16 *The behavior of* $\Theta^{[\mathfrak{j}]}_{\mathfrak{P},i}$ *on q-expansions.* Fix a prime $\mathfrak{P}$ of O_L over p and integers $1 \leq i \leq f_{\mathfrak{P}}$ and $0 \leq \mathfrak{j} \leq e_{\mathfrak{P}} - 1$. To compute the effect of $\Theta^{[\mathfrak{j}]}_{\mathfrak{P},i}$ on q-expansions we need to use some definitions and properties of Tate objects; see 6.3. Fix a cusp $(\mathfrak{A},\mathfrak{B},\varepsilon,\mathrm{j})$ as in 6.4. Fix a rational polyhedral cone decomposition $\{\sigma_\beta\}_\beta$ of the dual cone to $M^+_{\mathbf{R}} \subset M_{\mathbf{R}}$, where $M := \mathfrak{A}\mathfrak{B}$, as in 6.3. Then

$$R\big((\mathfrak{A},\mathfrak{B},\sigma_\beta)\big) := R \underset{\mathbf{Z}}{\otimes} \mathbf{Z}\big((\mathfrak{A},\mathfrak{B},\sigma_\beta)\big)$$

is a formally smooth R-algebra of dimension g, which can be interpreted as

$$\mathrm{Spec}\Big(R\big((\mathfrak{A},\mathfrak{B},\sigma_\beta)\big)\Big) = \left(S^\wedge_{\sigma_\beta} \backslash S_{\sigma_\beta,0}\right) \underset{\mathbf{Z}}{\times} R.$$

Moreover,

$$\Omega^1_{R((\mathfrak{A},\mathfrak{B},\sigma_\beta))/R} = \left\langle \frac{dq^\nu}{q^\nu} \right\rangle_{\nu \in M},$$

the span is taken as a $R\big((\mathfrak{A},\mathfrak{B},\sigma_\beta)\big)$-module. We conclude that

12.17 Lemma. *There is a canonical isomorphism*

$$\Omega^1_{R((\mathfrak{A},\mathfrak{B},\sigma_\beta))/R} \cong R\big((\mathfrak{A},\mathfrak{B},\sigma_\beta)\big) \underset{\mathbf{Z}}{\otimes} M.$$

In particular, the module of relative differentials $\Omega^1_{R((\mathfrak{A},\mathfrak{B},\sigma_\beta))/R}$ *is endowed with the structure of free* $R\big((\mathfrak{A},\mathfrak{B},\sigma_\beta)\big) \otimes_{\mathbf{Z}} O_L$*-module of rank 1.*

12.18 Remark. By 6.2 the translation invariant relative differentials

$$\omega_{\mathbf{Tate}(\mathfrak{A},\mathfrak{B})_{\sigma_\beta}/R\left((\mathfrak{A},\mathfrak{B},\sigma_\beta)\right)}$$

of the universal object $\mathbf{Tate}(\mathfrak{A},\mathfrak{B})_{\sigma_\beta}$ over $R\big((\mathfrak{A},\mathfrak{B},\sigma_\beta)\big)$ are canonically isomorphic to $R\big((\mathfrak{A},\mathfrak{B},\sigma_\beta)\big)\otimes_{\mathbf{Z}}\mathfrak{A}$ as $R\big((\mathfrak{A},\mathfrak{B},\sigma_\beta)\big)\otimes_{\mathbf{Z}} O_L$-module. The O_L-structure is of course given by the real multiplication by O_L on $\mathbf{Tate}(\mathfrak{A},\mathfrak{B})_{\sigma_\beta}$ over $R\big((\mathfrak{A},\mathfrak{B},\sigma_\beta)\big)$. In 12.4 we have fixed an element $l\in\mathfrak{I}=\mathrm{Hom}_{O_L}(\mathfrak{B},\mathfrak{A})$ inducing an O_L-linear isomorphism $\mathbf{Z}_p\otimes_{\mathbf{Z}}\mathfrak{B}\xrightarrow{\sim}\mathbf{Z}_p\otimes_{\mathbf{Z}}\mathfrak{A}$. We get an O_L-linear isomorphism

$$\tau\colon \mathbf{Z}_p\underset{\mathbf{Z}}{\otimes}(\mathfrak{A}\mathfrak{B})\xrightarrow{\sim}\mathbf{Z}_p\underset{\mathbf{Z}}{\otimes}\mathfrak{A}^2.$$

12.19 Proposition. *The Kodaira-Spencer map is defined on $R\big((\mathfrak{A},\mathfrak{B},\sigma_\beta)\big)$. The O_L-linear isomorphism τ makes the following diagram commutative*

$$\begin{array}{ccc}
\Omega^1_{R\left((\mathfrak{A},\mathfrak{B},\sigma_\beta)\right)/R} & \xrightarrow{\mathrm{KS}} & \omega^{\otimes^2_{O_L}}_{\mathbf{Tate}(\mathfrak{A},\mathfrak{B})_{\sigma_\beta}/R\left((\mathfrak{A},\mathfrak{B},\sigma_\beta)\right)} \\
\downarrow\wr & & \downarrow\wr \\
R\big((\mathfrak{A},\mathfrak{B},\sigma_\beta)\big)\otimes_{\mathbf{Z}} M & \xrightarrow{1\otimes\tau} & R\big((\mathfrak{A},\mathfrak{B},\sigma_\beta)\big)\otimes_{\mathbf{Z}}\mathfrak{A}^{\otimes^2_{O_L}},
\end{array}$$

Proof: It follows from [Ka4, §1.1.18-§1.1.20].

12.20 Remark. The isomorphisms KS and τ depend on the choice of the element $l\in\mathfrak{I}^+$ in 12.4. Different choices of l change the isomorphisms by multiplication by a unit of $R\otimes_{\mathbf{Z}} O_L$. In [Ka4, §1.0.21] one finds an expression for KS which is independent of such choice.

12.21 Corollary. *The notation is as in 12.12. Let $f\in\Gamma\big(\mathfrak{M}(m,n),O_{\mathfrak{M}(m,n)}\big)$ be a regular function on $\mathfrak{M}(m,n)$. Fix a $\mathfrak{I}$-polarized unramified cusp $(\mathfrak{A},\mathfrak{B},\varepsilon,\mathrm{j}_\varepsilon)$ of $\mathfrak{M}(m,n)$. Suppose that f is equal to $\sum_{\nu\in M^+\cup\{0\}}a_\nu q^\nu$ in $R/\mathfrak{m}^m\big((\mathfrak{A},\mathfrak{B},\sigma_\beta)\big)$. For any $\nu\in\mathfrak{P}^{\mathrm{j}}M$ define*

$$\tilde\chi^{[\mathrm{j}]}_{\mathfrak{P},i}(\nu)=\tilde\sigma^{[\mathrm{j}]}_{\mathfrak{P},i}\big(\mathrm{j}_\varepsilon(\tau(\nu))\big)\bmod \mathfrak{m}^{t^{[\mathrm{j}]}_{\mathfrak{P}}(m)},$$

where $\tilde\sigma^{[\mathrm{j}]}_{\mathfrak{P},i}\colon\big[R\otimes_{\mathbf{Z}}\mathfrak{P}^{\mathrm{j}}\big]\to R$ is the $R\otimes_{\mathbf{Z}}O_L$-linear homomorphism defined in 12.8. Then

1. *$f\in\Gamma\big(\mathfrak{M}(m,n),O^{\mathfrak{P},[\mathrm{j}]}_{\mathfrak{M}(m,n)}\big)$ in the sense of 12.12 if and only if $\nu\in\mathfrak{P}^{\mathrm{j}}M$ for all ν such that $a_\nu\neq 0$;*
2. *if f is in $\Gamma\big(\mathfrak{M}(m,n),O^{\mathfrak{P},[\mathrm{j}]}_{\mathfrak{M}(m,n)}\big)$, the value of $\Theta^{[\mathrm{j}]}_{\mathfrak{P},i}(f)$ in $R/\mathfrak{m}^{t^{[\mathrm{j}](m)}_{\mathfrak{P}}}\big((\mathfrak{A},\mathfrak{B},\sigma_\beta)\big)$ is*

$$\Theta^{[\mathrm{j}]}_{\mathfrak{P},i}(f)=\sum_\nu\tilde\chi^{[\mathrm{j}]}_{\mathfrak{P},i}(\nu)a_\nu q^\nu.$$

Proof: Since the condition $f \in \Gamma\big(\mathfrak{M}(m,n), O_{\mathfrak{M}(m,n)}^{\mathfrak{P},[\mathrm{j}]}\big)$ is a closed condition, it is enough to check it at the cusp $(\mathfrak{A},\mathfrak{B},\varepsilon,\mathrm{j}_\varepsilon)$. Note that

$$\begin{aligned} d\left(a_0+\sum_{\nu\in M^+} a_\nu q^\nu\right) &= \sum_{\nu\in M^+} a_\nu q^\nu \frac{d(q^\nu)}{q^\nu} \\ &= \sum_{\nu\in M^+} a_\nu q^\nu \otimes \nu, \end{aligned}$$

where the last element is in $R\big((\mathfrak{A},\mathfrak{B},\sigma_\beta)\big)\otimes_{\mathbf{Z}} M$. By 12.19 we have that f belongs to $\Gamma\big(\mathfrak{M}(m,n), O_{\mathfrak{M}(m,n)}^{\mathfrak{P},[\mathrm{j}]}\big)$ if and only if $\sum_{\nu\in M^+} a_\nu q^\nu\otimes\nu$ is in the image of $R\big((\mathfrak{A},\mathfrak{B},\sigma_\beta)\big)\otimes_{\mathbf{Z}} \mathfrak{P}^{\mathrm{j}} M$ in $R\big((\mathfrak{A},\mathfrak{B},\sigma_\beta)\big)\otimes_{\mathbf{Z}} M$. This proves (I).

By 6.5 the element $\omega^{\mathrm{can}}\otimes\omega^{\mathrm{can}}$ in $\omega^{\otimes^2_{O_L}}_{\mathbf{Tate}(\mathfrak{A},\mathfrak{B})_{\sigma_\beta}/R((\mathfrak{A},\mathfrak{B},\sigma_\beta))}$ is the inverse image of 1 via the composite homomorphism called, say, J

$$\omega^{\otimes^2_{O_L}}_{\mathbf{Tate}(\mathfrak{A},\mathfrak{B})_{\sigma_\beta}/R((\mathfrak{A},\mathfrak{B},\sigma_\beta))} \longrightarrow R\big((\mathfrak{A},\mathfrak{B},\sigma_\beta)\big)\underset{\mathbf{Z}}{\otimes}\mathfrak{A}^{\otimes^2_{O_L}} \xrightarrow{1\otimes \mathrm{j}_\varepsilon} R\big((\mathfrak{A},\mathfrak{B},\sigma_\beta)\big)\underset{\mathbf{Z}}{\otimes} O_L.$$

By the definition of $\Theta^{[\mathrm{j}]}_{\mathfrak{P},i}$ in 12.12 this implies that $\Theta^{[\mathrm{j}]}_{\mathfrak{P},i}(f)\big(\mathbf{Tate}(\mathfrak{A},\mathfrak{B}),\varepsilon,\mathrm{j}_\varepsilon\big) = \tilde{\sigma}^{[\mathrm{j}]}_{\mathfrak{P},i}\left(J\big(\mathrm{KS}(df)\big)\cdot\omega^{\mathrm{can}}\otimes\omega^{\mathrm{can}}\right) = \tilde{\sigma}^{[\mathrm{j}]}_{\mathfrak{P},i}\left(J\big(\mathrm{KS}(df)\big)\right)$ (the first $\tilde{\sigma}^{[\mathrm{j}]}_{\mathfrak{P},i}$ is defined on differentials as in 12.12, the second $\tilde{\sigma}^{[\mathrm{j}]}_{\mathfrak{P},i}$ is defined on $\big[R\big((\mathfrak{A},\mathfrak{B},\sigma_\beta)\big)\otimes_{\mathbf{Z}}\mathfrak{P}^{\mathrm{j}}\big]$). Since the q-expansion of the modular form $a\big(\chi_{\mathfrak{P},i}\big)$ at any unramified cusp is 1, we have that the q-expansion of $\Theta^{[\mathrm{j}]}_{\mathfrak{P},i}(f)$ at the given cusp is

$$\begin{aligned} \Theta^{[\mathrm{j}]}_{\mathfrak{P},i}(f)\big(\mathbf{Tate}(\mathfrak{A},\mathfrak{B}),\varepsilon,\mathrm{j}_\varepsilon\big) &= \tilde{\sigma}^{[\mathrm{j}]}_{\mathfrak{P},i}\left(J\Big(\mathrm{KS}\big(d(a_0+\sum_{\nu\in M^+} a_\nu q^\nu)\big)\Big)\right) \\ &= \sum_{\nu\in M^+} a_\nu q^\nu\cdot\tilde{\sigma}^{[\mathrm{j}]}_{\mathfrak{P},i}\Big(\mathrm{j}_\varepsilon\big(\tau(\nu)\big)\Big) \end{aligned}$$

as claimed.

12.22 Corollary. *The operators $\Theta^{[\mathrm{j}]}_{\mathfrak{P},i}$ commute for different primes $\mathfrak{P}$ and different $1\le i\le f_{\mathfrak{P}}$.*

12.23 Corollary. *The notation is as in 12.15 and in 12.21. Let f be a $\mathfrak{I}$-polarized p-adic Hilbert modular form over R of level μ_N and weight χ; see 11.4. Suppose that f has q-expansion at the $\mathfrak{I}$-polarized p-adic cusp $\big(\mathfrak{A},\mathfrak{B},\varepsilon_{p^\infty N},\mathrm{j}_\varepsilon\big)$ equal to $a_0+\sum_{\nu\in M^+} a_\nu q^\nu$; see 11.6 for the notation. Here $M=\mathfrak{A}\mathfrak{B}$. Then*

1. *$f\in\mathbf{M}(R,\mu_N,\chi)^{p-\mathrm{adic},\mathfrak{P},[\mathrm{j}]}$ if and only if $a_\nu=0$ for all $\nu\notin\mathfrak{P}^{\mathrm{j}} M$;*
2. *if $f\in\mathbf{M}(R,\mu_N,\chi)^{p-\mathrm{adic},\mathfrak{P},[\mathrm{j}]}$, the q-expansion of $\Theta^{[\mathrm{j}]}_{\mathfrak{P},i}(f)$ at the same cusp is*

$$\Theta^{[\mathrm{j}]}_{\mathfrak{P},i}(f)\big(\mathbf{Tate}(\mathfrak{A},\mathfrak{B}),\varepsilon,\mathrm{j}_\varepsilon\big) = \sum_\nu \tilde{\chi}^{[\mathrm{j}]}_{\mathfrak{P},i}(\nu) a_\nu q^\nu.$$

Proof: It follows from the definition of $\Theta^{[\mathrm{j}]}_{\mathfrak{P},i}$ given in 12.15 and the previous corollary.

12.24 Corollary. *The p-adic theta operators $\Theta_{\mathfrak{P},i}^{[j]}$ commute for different primes $\mathfrak{P}$ and different $1 \leq i \leq f_{\mathfrak{P}}$.*

12.25 *The comparison with the complex theory.* We use the notation of 6.11. Given a class in the strict class group of L, we may choose a representative $\mathfrak{I}$ which is fractional ideal prime to p. Take $\mathfrak{B} = O_L$ and $\mathfrak{A} = \mathfrak{I}$. Under the above assumption there is a canonical identification $j_{\text{can}}: \mathfrak{A} \otimes_{\mathbf{Z}} \mathbf{Z}_p \xrightarrow{\sim} O_L \otimes_{\mathbf{Z}} \mathbf{Z}_p$ and, hence, canonical isomorphisms $\varepsilon_{p^n}: p^{-n}O_L/O_L \xrightarrow{\sim} p^{-n}\mathfrak{A}/\mathfrak{A}$ for every $n \in \mathbf{N}$. This defines a canonical $\mathfrak{I}$-polarized unramified cusp $(\mathfrak{A}, \mathfrak{B}, \varepsilon_{p^\infty N}, j_\varepsilon)$ with $j_\varepsilon = j_{\text{can}}$; see 6.5 and 11.6.

Let f be a $\mathfrak{I}$-polarized Hilbert modular forms in $\mathbf{M}(\bar{\mathbf{Q}}, \mu_N, \chi)$. Choose embeddings $\bar{\mathbf{Q}} \subset \mathbf{C}$ and $\bar{\mathbf{Q}} \subset \bar{\mathbf{Q}}_p$ and view f as complex or p-adic Hilbert modular form. Assume that the q-expansion of f at the cusp $(i\infty, \ldots, i\infty)$ is $a_0 + \sum_{\nu \in \mathfrak{A}^+} a_\nu q^\nu$. Then, $f \in \mathbf{M}(\bar{\mathbf{Q}}, \mu_N, \chi)^{\mathfrak{P},[j]}$ if and only if $a_\nu = 0$ for all $\nu \notin (\mathfrak{P}^j\mathfrak{A})^+$ and

$$\Theta_{\mathfrak{P},i}^{[j]}(f)\big(\mathbf{Tate}(\mathfrak{A}, \mathfrak{B}), \varepsilon_{p^\infty N}, j_\varepsilon\big) = \sum_{\nu \in \mathfrak{A}^+} \chi_{\mathfrak{P},i}\big(\nu \pi_{\mathfrak{P}}^{-j}\big) a_\nu q^\nu;$$

where $\pi_{\mathfrak{P}}$ is the chosen uniformizer of O_L at p.

12.26 *Katz's p-adic theta operators.* In [Ka4, §2.6] one finds a definition of p-adic theta operators on p-adic Hilbert modular forms à la Katz (see 11.4) if p *is unramified* in L. In this case it follows from [Ka4, Cor. 2.6.25] and 12.23 that Katz's theta operators coincide on q-expansions with the p-adic theta operators defined above.

12.27 *Other examples of p-adic modular forms.* One way to produce examples of p-adic modular forms (see 11.4) is by applying the p-adic theta operators to classical Hilbert modular forms. In general, the image of a classical modular form, i. e., an element of $M(F, \mu_N, \chi)$ (see 5.1), under a theta operator is not a classical modular form. To illustrate that we consider the case of $g = 2$ and the $\mathfrak{I}$-polarized classical Eisenstein series E_2 over $\mathbf{Q}$ of weight 2. Recall from 18.3 that its q-expansion at a $\mathfrak{I}$-polarized unramified $\mathbf{Q}$-rational cusp $(\mathfrak{A}, \mathfrak{B}, j_{\text{can}})$ is

$$\mathbf{Nm}^{k-1}(\mathfrak{A})\left(2^{-g}\zeta_L(1-k) + \sum_{\nu \in (\mathfrak{A}\mathfrak{B})^+}\Big(\sum_{\nu \in \mathfrak{C} \subset \mathfrak{A}\mathfrak{B}} \mathbf{Nm}(\nu\mathfrak{C}^{-1})^{k-1}\Big)q^\nu\right).$$

Let p be a split prime and let χ, χ' be the basic characters over p. Then $\Theta_\chi E_2$ is a p-adic modular form, whose q-expansion is

$$\mathbf{Nm}^{k-1}(\mathfrak{A})\left(2^{-g}\zeta_L(1-k) + \sum_{\nu \in (\mathfrak{A}\mathfrak{B})^+}\Big(\sum_{\nu \in \mathfrak{C} \subset \mathfrak{A}\mathfrak{B}} \mathbf{Nm}(\nu\mathfrak{C}^{-1})^{k-1}\Big)\tilde{\chi}(\nu)q^\nu\right).,$$

which is not a classical modular form. To see that we remark that there exists a pull-back map from classical modular forms for our quadratic field to modular forms on $\mathbf{Q}$, taking a modular form of level 1 (say) and weight $\chi_1^{a_1} \ldots \chi_g^{a_g}$ to a modular form on $\mathrm{SL}_2(\mathbf{Z})$ of weight $a_1 + \ldots + a_g$. See § 17. The pull-back of E_2 is a multiple of the Eisenstein series $\mathrm{E}_4^{\mathbf{Q}}$ on $\mathrm{SL}_2(\mathbf{Z})$, where to avoid confusion, we write $\mathrm{E}_k^{\mathbf{Q}}$ for Eisenstein series of weight k for the field $\mathbf{Q}$.

If $\Theta_\chi E_2$ is classical of some weight and level, so is its Galois conjugate $\Theta_{\chi'} E_2$ over L. Hence, the sum $\Theta_\chi E_2 + \Theta_{\chi'} E_2$ is in the graded ring of classical modular forms. It follows from 17.8 that the pull-back of $\Theta_\chi E_2 + \Theta_{\chi'} E_2$ is proportional to $\Theta \mathrm{E}_4^{\mathbf{Q}}$ and is a p-adic cusp form on $\mathrm{SL}_2(\mathbf{Q})$ of weight 6 and integral q-expansion. But then, reducing modulo p, we would have a mod p cusp form of weight $4+p+1$, hence divisible, as a holomorphic modular form, by Δ. Take $p = 2, 3$ or 5 to obtain a contradiction.

From now on we work in characteristic p. The goal of the rest of this section is to define theta operators on Hilbert modular forms in characteristic p. See the introduction of the section for a more detailed discussion.

12.28 *The poles of $\Theta_{\mathfrak{P},i}$ in characteristic p.* We use the notation of 11.1 and define

$$\mathfrak{M} := \mathfrak{M}(1,1)^{\mathrm{Kum}}, \qquad \mathfrak{M}' := \mathfrak{M}(1,0).$$

Then $\phi\colon \mathfrak{M} \to \mathfrak{M}'$ is a Galois cover with group

$$G \cong \prod_{\mathfrak{P}|(p)} (O_L/\mathfrak{P})^*.$$

By 9.1 and 9.3 we can complete ϕ to a finite morphism of quasi-projective normal schemes:

$$\begin{array}{ccccc} \mathfrak{M} & \hookrightarrow & \mathfrak{M}^* & \hookrightarrow & \overline{\mathfrak{M}}(k,\mu_{pN})^{\mathrm{Kum}} \\ \downarrow{\scriptstyle \phi} & & \downarrow{\scriptstyle \overline{\phi}} & & \downarrow{\scriptstyle \overline{\phi}} \\ \mathfrak{M}' & \hookrightarrow & \mathfrak{M}(k,\mu_N)^{\mathrm{R}} & \hookrightarrow & \overline{\mathfrak{M}}(k,\mu_N). \end{array}$$

Note that $\mathfrak{M}^*$ has codimension at least 2 in $\overline{\mathfrak{M}}(k,\mu_{pN})^{\mathrm{Kum}}$ and the map $\overline{\phi}$ is ramified along the complement of the ordinary locus of $\overline{\mathfrak{M}}(k,\mu_N)$.

Let $\pi\colon \mathrm{A}' \to \mathfrak{M}(k,\mu_N)$ be the universal abelian scheme. Proceeding as in 12.6 one obtains an isomorphism

$$\mathrm{KS}'\colon \Omega^1_{\mathfrak{M}(k,\mu_N)^{\mathrm{R}}/k} \xrightarrow{\sim} \omega^{\otimes^2_{O_L}}_{\mathrm{A}'/\mathfrak{M}(k,\mu_N)^{\mathrm{R}}},$$

which we use to define an isomorphism

$$\mathrm{KS} := \overline{\phi}^*\left(\mathrm{KS}'\right)\colon \overline{\phi}^*\left(\Omega^1_{\mathfrak{M}(k,\mu_N)^{\mathrm{R}}/k}\right) \xrightarrow{\sim} \overline{\phi}^*\left(\omega^{\otimes^2_{O_L}}_{\mathrm{A}'/\mathfrak{M}(k,\mu_N)^{\mathrm{R}}}\right) = \omega^{\otimes^2_{O_L}}_{\overline{\phi}^*(\mathrm{A}')/\mathfrak{M}^*},$$

extending to $\mathfrak{M}^*$ the previously defined KS on $\mathfrak{M}$.

12.29 Definition. *For every prime $\mathfrak{P}$ over p and any integer $1 \le i \le f_{\mathfrak{P}}$, define the Weil divisor of $\mathfrak{M}^*$:*

$$W_{\mathfrak{P},i} := \text{support of the effective divisor } \overline{\phi}^{-1}\left(h_{\mathfrak{P},i}\right).$$

12.30 Remark. By construction $W_{\mathfrak{P},i}$ is reduced. Moreover,

$$\left(\overline{\phi}^{-1}(h_{\mathfrak{P},i})\right) = \overline{\phi}^{-1}\left((h_{\mathfrak{P},i})\right) = \left(p^{f_{\mathfrak{P}}} - 1\right) W_{\mathfrak{P},i}.$$

See 9.6.

If f is a regular function of $\mathfrak{M}$ lying in $\Gamma\left(\mathfrak{M}, O_{\mathfrak{M}}^{\mathfrak{P},[j]}\right)$, we are interested in computing the poles of $\Theta^{[j]}_{\mathfrak{P},i}(f)$ on $\mathfrak{M}^*$. This is achieved as follows:

12.31 *Some notation.* Let $\mathfrak{P}$ and $\mathfrak{Q}$ be prime ideals of O_L over p and let $1 \leq i \leq f_{\mathfrak{P}}$ and $1 \leq j \leq f_{\mathfrak{Q}}$. Define

$$\left(\Omega^1_{\mathfrak{M}/k}\right)_{\mathfrak{P},i} := e_{\mathfrak{P},i} \cdot \Omega^1_{\mathfrak{M}/k}$$

and if $f \in O_{\mathfrak{M}}$ define

$$df_{\mathfrak{P},i} := e_{\mathfrak{P},i} \cdot df \in \left(\Omega^1_{\mathfrak{M}/k}\right)_{\mathfrak{P},i}.$$

See 12.7 for the definition of the idempotent $e_{\mathfrak{P},i}$. Let δ be a local uniformizer of an irreducible component of the divisor $W_{\mathfrak{Q},j}$. Let v_δ be the discrete valuation on the meromorphic functions on $\mathfrak{M}$ defined by δ. Let us write

$$f = \frac{u}{\delta^n},$$

where u is a function such that $v_\delta(u) = 0$. Then

$$\left(\Omega^1_{\mathfrak{M}/k}\right)_{\mathfrak{P},i} \ni (df)_{\mathfrak{P},i} = \frac{(du)_{\mathfrak{P},i}}{\delta^n} - \frac{nu(d\delta)_{\mathfrak{P},i}}{\delta^{n+1}}. \tag{12.31.1}$$

12.32 *The poles of $d\delta$ for a specific choice of δ.* The modular forms $h_{\mathfrak{P},i}$ are all well defined functions on $\mathfrak{M}^*$ via the trivialization of the line bundles $\mathcal{L}_{\chi_{\mathfrak{P},i}}$ by means of the sections $a(\chi_{\mathfrak{P},i})$ (note that this trivialization exists only over $\mathfrak{M}$, but that suffices for viewing the $h_{\mathfrak{P},i}$'s as functions). We remark that the value of the function $h_{\mathfrak{P},i}$ on a geometric point $[(A, \iota, \lambda, \varepsilon_{pN})]$ of $\mathfrak{M}$ is the value of the modular form $h_{\mathfrak{P},i}$ on $\left[(A, \iota, \lambda, \varepsilon_N, \varepsilon_{p*}(dt/t))\right]$. By 9.3-9.4 we may choose $\delta = \delta(\mathfrak{Q}, j)$ such that

$$\delta^{p^{f_{\mathfrak{Q}}}-1} = h^{p^{f_{\mathfrak{Q}}-1}}_{\mathfrak{Q},j+1} \cdots h^p_{\mathfrak{Q},j-1} h_{\mathfrak{Q},j}.$$

Since k is of characteristic p, we conclude that

$$(p^{f_{\mathfrak{Q}}} - 1)\delta^{p^{f_{\mathfrak{Q}}}-2} d\delta = \left(h^{p^{f_{\mathfrak{Q}}-1}}_{\mathfrak{Q},j+1} \cdots h^p_{\mathfrak{Q},j-1}\right) dh_{\mathfrak{Q},j}. \tag{12.32.1}$$

12.33 Proposition. *We have* $dh_{\mathfrak{Q},j} \in \left(\Omega^1_{\mathfrak{M}/k}\right)_{\mathfrak{Q},j}$.

Proof: The statement is equivalent to say that the $(\mathfrak{P}, i)$-projection of $dh_{\mathfrak{Q},j}$ is zero for $(\mathfrak{P}, i) \neq (\mathfrak{Q}, j)$. Therefore, it suffices to prove the Proposition after reducing modulo the maximal ideal of every geometric point of $\mathfrak{M}$.

Fix a geometric point $[(A_0, \iota_0, \lambda_0, (\varepsilon_0)_{pN})]$. Let $k[\![t^{[j]}_{\mathfrak{P},i}]\!]_{\mathfrak{P},i,j}$ be the completion of $\mathfrak{M}'$ at $[(A_0, \iota_0, \lambda_0, (\varepsilon_0)_N)]$ as in 8.11. Let $\mathfrak{m}_\iota$ be its maximal ideal and let

$$R_\iota := k[\![t^{[j]}_{\mathfrak{P},i}]\!]_{\mathfrak{P},i,j}/\mathfrak{m}_\iota^2.$$

By abuse of notation, $\mathfrak{m}_\iota$ denotes the maximal ideal of R_ι.

With the notation of 8.15, it follows from 8.17 that in $\mathfrak{m}_\iota/\mathfrak{m}_\iota^2$

$$dh_{\mathfrak{Q},j} = \bar{c}^{[1]}_{\mathfrak{Q},j} dt^{[1]}_{\mathfrak{Q},j} \tag{12.33.1}$$

for an invertible element $\bar{c}^{[1]}_{\mathfrak{Q},j}$ of k. The Proposition follows from the following:

12.34 Lemma. *For every geometric point* $[(A_0, \iota_0, \lambda_0, (\varepsilon_0)_N)]$ *of* $\mathfrak{M}(k, \mu_N)^{\mathrm{R}}$ *one has*

$$k dt^{[1]}_{\mathfrak{Q},j} \oplus k\pi_{\mathfrak{Q}} dt^{[1]}_{\mathfrak{Q},j} \oplus \ldots \oplus k\pi_{\mathfrak{Q}}^{e_Q-1} dt^{[1]}_{\mathfrak{Q},j} = \left(\Omega^1_{R_\iota/k}\right)_{\mathfrak{Q},j}.$$

Proof: The proof relies on understanding the connection between two deformation theories of abelian varieties: one based on Grothendieck and Mumford's infinitesimal theory, the other based on the theory of displays. Consider the display $(\mathrm{P}_0, \mathrm{Q}_0, \mathrm{F}_0, \mathrm{V}_0^{-1})$ over k introduced in 8.1–8.3. The following results follow from [Zi].

a) Let $(\mathrm{P}_1, \mathrm{Q}_1, \mathrm{F}_1, \mathrm{V}_1^{-1})$ and $(\mathrm{P}_2, \mathrm{Q}_2, \mathrm{F}_2, \mathrm{V}_2^{-1})$ be two displays over an artinian k-algebra D deforming $(\mathrm{P}_0, \mathrm{Q}_0, \mathrm{F}_0, \mathrm{V}_0^{-1})$. For $i = 1, 2$ let $\widehat{\mathrm{Q}}_i$ be the inverse image of Q_i via $\mathrm{P} \to \mathrm{P}_0$. Then there exists a unique isomorphism

$$\alpha_{\mathrm{P}_1,\mathrm{P}_2}\colon (\mathrm{P}_1, \widehat{\mathrm{Q}}_1, \mathrm{F}_1, \mathrm{V}_1^{-1}) \xrightarrow{\sim} (\mathrm{P}_2, \widehat{\mathrm{Q}}_2, \mathrm{F}_1, \mathrm{V}_2^{-1})$$

inducing the identity on $(\mathrm{P}_0, \mathrm{Q}_0, \mathrm{F}_0, \mathrm{V}_0^{-1})$. In particular, the functor associating to a nilpotent pd-thickening $k \subset D$ the D-module $\mathcal{D}_{\mathrm{P}_0}(D) := \bar{\mathrm{P}} := \mathrm{P}/I_D\mathrm{P}$, where $(\mathrm{P}, \mathrm{Q}, \mathrm{F}, \mathrm{V}^{-1})$ is a display over D deforming $(\mathrm{P}_0, \mathrm{Q}_0, \mathrm{F}_0, \mathrm{V}_0^{-1})$, defines a crystal over k.

b) Let $\mathfrak{D}_{A_0[p^\infty]}$ be the crystal over k associated to the formal group $A_0[p^\infty]$ by Grothendieck and Messing [Me]. The morphism $\mathcal{D}_{\mathrm{P}_0} \to \mathfrak{D}_{A_0[p^\infty]}$, which associates to a nilpotent pd-thickening $k \subset D$ and a display $(\mathrm{P}, \mathrm{Q}, \mathrm{F}, \mathrm{V}^{-1})$ over D deforming $(\mathrm{P}_0, \mathrm{Q}_0, \mathrm{F}_0, \mathrm{V}_0^{-1})$ the Lie algebra of the universal extension over D of the formal group associated to P, defines an isomorphism of crystals; see [Zi, Thm. 6].

Let D be an artinian k-algebra endowed with nilpotent divided powers structure and a section s. Denote by $(\mathrm{P}_{\mathrm{can}}, \mathrm{Q}_{\mathrm{can}}, \mathrm{F}_{\mathrm{can}}, \mathrm{V}_{\mathrm{can}}^{-1})$ the trivial deformation of $(\mathrm{P}_0, \mathrm{Q}_0, \mathrm{F}_0, \mathrm{V}_0^{-1})$ to D defined by pulling-back via s. Let $\mathrm{Def}_{\mathrm{P}_0}(D)$ be the category of deformations of the display $(\mathrm{P}_0, \mathrm{Q}_0, \mathrm{F}_0, \mathrm{V}_0^{-1})$ to a display over D. Let $\mathrm{Def}_{\bar{\mathrm{Q}} \subset \bar{\mathrm{P}}_{\mathrm{can}}}(D)$ be the category of liftings of the Hodge filtration $\bar{\mathrm{Q}}_0 \subset \bar{\mathrm{P}}_0$ in $\bar{\mathrm{P}}_{\mathrm{can}}$.

c) The functor

$$\mathrm{Def}_{\mathrm{P}_0}(D) \longrightarrow \mathrm{Def}_{\bar{\mathrm{Q}} \subset \bar{\mathrm{P}}_{\mathrm{can}}}(D),$$

associating to $(\mathrm{P}, \mathrm{Q}, \mathrm{F}, \mathrm{V}^{-1}) \in \mathrm{Def}_{\mathrm{P}_0}(D)$ the flag $\alpha_{\mathrm{P}_{\mathrm{can}},\mathrm{P}}^{-1}(\bar{\mathrm{Q}}) \subset \bar{\mathrm{P}}_{\mathrm{can}}$, defines an equivalence of categories.

d) Let $R := k[\![t_{a,b}]\!]_{1 \le a,b \le g}$ with the relations $t_{a,b}t_{a',b'} = 0$. Let T (resp. $\bar{T}$) be the matrix of Teichmüller lifts $(w(t_{a,b}))_{1 \le a,b \le g}$ (resp. the matrix $(t_{a,b})_{1 \le a,b \le g}$). The flag $\bar{\mathrm{Q}}_{\mathrm{can}} + \bar{T}\bar{\mathfrak{T}}_{\mathrm{can}} \subset \bar{\mathrm{P}}_{\mathrm{can}}$ comes from the display $\mathrm{P} := \mathrm{P}_{\mathrm{can}}$ and $\mathrm{Q} := \mathrm{Q}_{\mathrm{can}}$ via the map

$$\alpha_{\mathrm{P}_{\mathrm{can}},\mathrm{P}} := \begin{pmatrix} \mathrm{Id}_g & T \\ 0 & \mathrm{Id}_g \end{pmatrix};$$

the matrix is given on $\mathrm{P}_{\mathrm{can}} = \mathfrak{T}_{\mathrm{can}} \oplus \mathfrak{L}_{\mathrm{can}}$ with respect to $\mathcal{B}$. In particular, the matrix of $\mathrm{F} \oplus \mathrm{V}^{-1}$ with respect to the same basis is given by

$$\begin{pmatrix} A + TC & B + TD \\ C & D \end{pmatrix}.$$

Imposing the condition that $(\mathrm{P}, \mathrm{Q}, \mathrm{F}, \mathrm{V}^{-1})$ is polarized is equivalent to restrict to the quotient of R defined by the relations $t_{a,b} = t_{b,a}$. The ring R_ι is the maximal k-algebra over which the O_L-action extends, see 8.11, and $\mathfrak{m}_\iota = (t_{a,b})$. Define $(\mathrm{P}, \mathrm{Q}, \mathrm{F}, \mathrm{V}^{-1})$ to be the associated display over R_ι.

Let $A \to \mathrm{Spec}(R_\iota)$ be the abelian scheme with real multiplication by O_L associated to the display $(\mathrm{P}, \mathrm{Q}, \mathrm{F}, \mathrm{V}^{-1})$. The isomorphism $\alpha_{\mathrm{P}_{\mathrm{can}},\mathrm{P}}$ induces an O_L-linear isomorphism

$$\mathrm{H}_{1,\mathrm{dR}}(A_0/k) \times_k \mathrm{Spec}(R_\iota) = \bar{P}_{\mathrm{can}} \xrightarrow{\sim} \bar{P} = \mathrm{H}_{1,\mathrm{dR}}(A/R_\iota). \tag{12.34.1}$$

By b) it is compatible with the isomorphism

$$\mathfrak{D}_{A_0[p^\infty]}(k) \times_k \mathrm{Spec}(R_\iota) \xrightarrow{\sim} \mathfrak{D}_{A_0[p^\infty]}(R_\iota).$$

By the comparison theorem between the crystals $\mathfrak{D}_{A_0[p^\infty]}$ and $R^1\pi_{\mathrm{crys},*}\big(O_{A_0,\mathrm{crys}}\big)$, we get that the isomorphism (12.34.1) identifies $\mathrm{H}_{1,\mathrm{dR}}(A_0/k)$ with the horizontal sections of the Gauss-Manin connection on $\mathrm{H}_{1,\mathrm{dR}}(A/R_\iota)$.

Via the identifications of 8.15, we get that the O_L-linear map deduced from (12.34.1)

$$\mathrm{Hom}\Big(\mathrm{H}^1\big(A_0, O_{A_0}\big), k\Big) = \bar{Q}_0 \longrightarrow \mathfrak{m}_\iota \cdot \big(\bar{P}/\bar{Q}\big) = \mathrm{Hom}\Big(\mathrm{H}^0\big(A_0, \Omega^1_{A_0/k}\big), \mathfrak{m}_\iota\Big)$$

is defined by the matrix $\bar{T}$. By [Ka4, §1.0.11–§1.0.21] the induced k-linear map

$$\mathrm{H}^0\big(A_0, \Omega^1_{A_0/k}\big) \otimes_{O_L} \mathrm{H}^0\big(A_0, \Omega^1_{A_0/k}\big) \xrightarrow{\sim} \mathfrak{m}_\iota = \Omega^1_{R_\iota/k}$$

coincides with the Kodaira-Spencer map. In particular, for any prime $\mathfrak{Q}$, any $1 \leq j \leq f_{\mathfrak{Q}}$ and any $1 \leq l \leq e_{\mathfrak{Q}} - 1$, we compute that

$$\big(\omega^{[1]}_{\mathfrak{Q},j}\big) \otimes \big(\omega^{[l]}_{\mathfrak{Q},j}\big) \longmapsto d\left(t^{[l]}_{\mathfrak{Q},j}\right)$$

by considering the corresponding matrix coefficient of $\bar{T}$. This implies that $dt^{[l]}_{\mathfrak{Q},j} = \pi^l_{\mathfrak{Q}} t^{[1]}_{\mathfrak{Q},j}$ for $1 \leq l \leq e_{\mathfrak{Q}} - 1$. This proves the lemma.

We are now ready to calculate the poles of $(d\delta)_{\mathfrak{P},i}$ *in the following sense*: we consider $(d\delta)_{\mathfrak{P},i}$ as a meromorphic section of the vector bundle $\overline{\phi}^*\big(\Omega^1_{\mathfrak{M}(k,\mu_N)^{\mathrm{R}}/k}\big)_{\mathfrak{P},i}$. As such it has a well defined divisor on $\mathfrak{M}^*$ (which is different from the divisor of $(d\delta)_{\mathfrak{P},i}$ considered as a meromorphic differential on $\mathfrak{M}^*$), if s is a meromorphic section and x is a point of height 1 with uniformizer t then the valuation of s at x is the maximal n such that s/t^n is a regular section.

By 8.17 and 12.34 the differentials $\{dh_{\mathfrak{Q},j}, \pi_{\mathfrak{Q}} dh_{\mathfrak{Q},j}, \ldots, \pi_{\mathfrak{Q}}^{e_{\mathfrak{Q}}-1} dh_{\mathfrak{Q},j}\}$ are local generators of $\big(\Omega^1_{\mathfrak{M}(k,\mu_N)^{\mathrm{R}}/k}\big)_{\mathfrak{Q},j}$ at $W_{\mathfrak{Q},j}$ as $O_{\mathfrak{M}(k,\mu_N)^{\mathrm{R}}}$-module. Gathering the results of 12.31 and 12.32 and using the Proposition, we conclude that for the particular choice of $\delta = \delta(\mathfrak{Q}, j)$ made in 12.32

$$v_\delta\Big(\big(d\delta\big)_{\mathfrak{P},i}\Big) = \begin{cases} \infty & \text{if } (\mathfrak{P}, i) \neq (\mathfrak{Q}, j); \\ 2 - p^{f_{\mathfrak{P}}} & \text{if } (\mathfrak{P}, i) = (\mathfrak{Q}, j). \end{cases}$$

Furthermore, $v_\delta\Big(\pi^l_{\mathfrak{Q}}\big(d\delta\big)_{\mathfrak{Q},j}\Big) = 2 - p^{f_{\mathfrak{Q}}}$ for every $0 \leq l \leq e_{\mathfrak{Q}} - 1$.

12.35 Proposition. *With the notation of 12.31, we have*

$$v_\delta\big((df)_{\mathfrak{P},i}\big) \begin{cases} \geq v_\delta(f) & \text{if } (\mathfrak{P}, i) \neq (\mathfrak{Q}, j); \\ \geq v_\delta(f) - (p^{f_{\mathfrak{P}}} - 2) & \text{if } (\mathfrak{P}, i) = (\mathfrak{Q}, j) \text{ and } p | v_\delta(f); \\ = v_\delta(f) - (p^{f_{\mathfrak{P}}} - 1) & \text{if } (\mathfrak{P}, i) = (\mathfrak{Q}, j) \text{ and } p \nmid v_\delta(f). \end{cases}$$

If f is an eigenfunction for the group G, $(\mathfrak{P}, i) = (\mathfrak{Q}, j)$ and $p | v_\delta(f)$, then

$$v_\delta\big((df)_{\mathfrak{P},i}\big) \geq v_\delta(f).$$

Proof: It remains to calculate the contribution of $v_\delta\big((du)_{\mathfrak{P},i}\big)$. Let B^0 (resp. B) be the local ring of $\mathfrak{M}(k,\mu_N)$ (resp. of $\mathfrak{M}^*$) at the component of $W_{\mathfrak{Q},j}$ corresponding to δ. By 9.6 the extension $B^0 \subset B$ factors as $B^0 \subset B^{\mathrm{et}} \subset B$ where $B^0 \subset B^{\mathrm{et}}$ is étale and $B = B^{\mathrm{et}}[\delta]$. Write $u = \sum_{h=0}^{p^{f_{\mathfrak{P}}}-2} u_h\delta^h$ with $u_h \in B^{\mathrm{et}}$. Then, $(du)_{\mathfrak{P},i} = \sum_h \Big(\delta^h(du_h)_{\mathfrak{P},i} + hu_h\delta^{h-1}(d\delta)_{\mathfrak{P},i}\Big)$. Note that $\delta^h(du_h)_{\mathfrak{P},i}$ lies in $B\otimes_{B^{\mathrm{et}}}\Omega^1_{B^{\mathrm{et}}/k} = \overline{\phi}^*\Omega^1_{B^0/k}$ and, in particular, it has no poles. Hence, $v_\delta\big((du)_{\mathfrak{P},i}\big) \geq \inf\{0, v_\delta(d\delta)_{\mathfrak{P},i}\}$.

Assume that f is an eigenfunction with respect to a character χ. Let $I \subset G$ be the Galois group of B/B^{et}. Then, I acts via roots of unity on δ: identifying $I = k_{\mathfrak{P}}^*$ we have $[\alpha]\delta = \alpha^{-1}\delta$ for every $\alpha \in I$. Then, $\sum_h u_h\chi(\alpha)\delta^{h-n} = \chi(\alpha)f = [\alpha]f = \sum_h u_h[\alpha]\delta^{h-n} = \sum_h u_h\alpha^{n-h}\delta^{h-n}$. Hence, for all $0 \leq h \leq p^{f_{\mathfrak{P}}}-2$ we have $u_h(\alpha^{n-h}-\chi(\alpha)) = 0$. Taking α to be a primitive element, one concludes that there exists only one h such that $\alpha^{n-h}-\chi(\alpha) = 0$. In particular, for every $h' \neq h$ one has $u_{h'} = 0$. Since $v_\delta(u) = 0$ by assumption, $h = 0$ and $u \in B^{\mathrm{et}}$. Reasoning as before, one gets the conclusion. Note that it also follows that χ is the character $\alpha \mapsto \alpha^n$ on I.

12.36 Lemma. *Let f be a regular function on $\mathfrak{M}$ such that $f \in \Gamma(\mathfrak{M}, O_{\mathfrak{M}}^{\mathfrak{Q},[\mathfrak{j}]})$ for $\mathfrak{j} > 0$. Let $f = u/\delta^n$, as above. Then, $p|n$ and $u \in \Gamma(\mathfrak{M}, O_{\mathfrak{M}}^{\mathfrak{Q},[\mathfrak{j}]})$.*

Proof: By definition $(df)_{\mathfrak{Q},j} \in \pi_{\mathfrak{Q}}^{\mathfrak{j}}(\Omega^1_{\mathfrak{M}/k})_{\mathfrak{Q},j}$. Since $(d\delta)_{\mathfrak{Q},j} \notin \pi_{\mathfrak{Q}}^{\mathfrak{j}}(\Omega^1_{\mathfrak{M}/k})_{\mathfrak{Q},j}$ if $\mathfrak{j} > 0$, it follows from (12.31.1) that $p|n$ and $u \in \Gamma(\mathfrak{M}, O_{\mathfrak{M}}^{\mathfrak{Q},[\mathfrak{j}]})$.

12.37 Proposition. *The notation is as in 12.12. Fix primes $\mathfrak{P}$ and $\mathfrak{Q}$ over p and integers $1 \leq i \leq f_{\mathfrak{P}}$ and $1 \leq j \leq f_{\mathfrak{Q}}$. Let $0 \leq \mathfrak{j} \leq e_{\mathfrak{P}}-1$. Let f be a regular function on $\mathfrak{M}$ such that $f \in \Gamma(\mathfrak{M}, O_{\mathfrak{M}}^{\mathfrak{P},[\mathfrak{j}]})$. Let C be an irreducible component of the divisor $W_{\mathfrak{Q},j}$ defined in 12.29. Let v_C be the corresponding valuation. Then*

$$v_C\left(\Theta^{[\mathfrak{j}]}_{\mathfrak{P},i}(f)\right)\begin{cases} \geq v_C(f) & \text{if } \mathfrak{P}\neq\mathfrak{Q};\\ \geq v_C(f) - 2p^{f_{\mathfrak{P}}-r} & \text{if } \mathfrak{P}=\mathfrak{Q} \text{ and } i\neq j;\\ \geq v_C(f)-2-\left(p^{f_{\mathfrak{P}}}-2\right) & \text{if } \mathfrak{P}=\mathfrak{Q},\ i=j \text{ and } p|v_C(f);\\ = v_C(f)-2-\left(p^{f_{\mathfrak{P}}}-1\right) & \text{if } \mathfrak{P}=\mathfrak{Q},\ i=j \text{ and } p\nmid v_C(f);\end{cases}$$

where $r = j-i$ if $j > i$ and $r = f_{\mathfrak{P}}+j-i$ if $j < i$. As before, if f is an eigenfunction for the group G, $(\mathfrak{P},i) = (\mathfrak{Q},j)$ and $p|v_C(f)$, then

$$v_C\left(\Theta^{[\mathfrak{j}]}_{\mathfrak{P},i}(f)\right) \geq v_C(f) - 2.$$

Proof: By 12.12, the identity

$$\Theta^{[\mathfrak{j}]}_{\mathfrak{P},i}(f) = \tilde{\sigma}^{[\mathfrak{j}]}_{\mathfrak{P},i}\Big(\mathrm{KS}(df)\Big)\cdot a(\chi^2_{\mathfrak{P},i})^{-1}$$

holds on $\mathfrak{M}$. We first explain how to extend it to $\mathfrak{M}^*$.

The map $\tilde{\sigma}^{[\mathfrak{j}]}_{\mathfrak{P},i}$ defined in 12.10 over $\mathfrak{M}(k,\mu_N)^{\mathrm{ord}}$ extends to an $O_L\otimes_{\mathbf{Z}} k$-linear surjective map

$$\tilde{\sigma}^{[\mathfrak{j}]}_{\mathfrak{P},i}\colon \left[\omega^{\otimes^2_{O_L}}_{\mathrm{A}'/\mathfrak{M}(k,\mu_N)^{\mathrm{R}}}\underset{O_L}{\otimes}\mathfrak{P}^{\mathfrak{j}}\right]\longrightarrow \mathcal{L}_{\chi^2_{\mathfrak{P},i}}$$

over $\mathfrak{M}(k,\mu_N)^{\mathrm{R}}$. Hence we obtain an $O_L\otimes_{\mathbf{Z}} k$-linear map

$$\left[\Omega^1_{\mathfrak{M}(k,\mu_N)^{\mathrm{R}}/k}\underset{O_L}{\otimes}\mathfrak{P}^{\mathfrak{j}}\right]\xrightarrow{\ \mathrm{KS}'\ }\left[\omega^{\otimes^2_{O_L}}_{\mathrm{A}'/\mathfrak{M}(k,\mu_N)^{\mathrm{R}}}\underset{O_L}{\otimes}\mathfrak{P}^{\mathfrak{j}}\right]\xrightarrow{\ \tilde{\sigma}^{[\mathfrak{j}]}_{\mathfrak{P},i}\ }\mathcal{L}_{\chi^2_{\mathfrak{P},i}}.$$

Pulling-back, we get a map

$$\overline{\phi}^*\left[\Omega^1_{\mathfrak{M}(k,\mu_N)^{\mathrm{R}}/k}\underset{O_L}{\otimes}\mathfrak{P}^{\mathrm{j}}\right]\longrightarrow\overline{\phi}^*\left(\mathcal{L}_{\chi^2_{\mathfrak{P},i}}\right).$$

We also have a natural inclusion of sheaves of $O_{\mathfrak{M}^{\mathrm{R}}}$-modules

$$0\longrightarrow\overline{\phi}^*\left(\Omega^1_{\mathfrak{M}(k,\mu_N)^{\mathrm{R}}/k}\right)\longrightarrow\Omega^1_{\mathfrak{M}^{\mathrm{R}}/k}.$$

The cokernel defines the branch locus of $\overline{\phi}$; see 9.6. Recall that our goal is to compute the poles of $\Theta^{[\mathrm{j}]}_{\mathfrak{P},i}(f)$ where df is to be interpreted as a meromorphic section of $\overline{\phi}^*\left(\Omega^1_{\mathfrak{M}(k,\mu_N)^{\mathrm{R}}/k}\right)$. Now $\Theta^{[\mathrm{j}]}_{\mathfrak{P},i}(f)=\tilde{\sigma}^{[\mathrm{j}]}_{\mathfrak{P},i}\Big(\mathrm{KS}(df)\Big)\cdot a\left(\chi^2_{\mathfrak{P},i}\right)^{-1}$ makes sense over $\mathfrak{M}^*$.

By general principles, if $\tau\colon\mathcal{L}_1\to\mathcal{L}_2$ is a surjective map of locally free sheaves with a locally free kernel over a normal scheme and if s is a meromorphic section of $\mathcal{L}_1$, then the divisor of s is less or equal to the divisor of $\tau(s)$. In particular, the poles of $\tilde{\sigma}^{[\mathrm{j}]}_{\mathfrak{P},i}(\mathrm{KS}(df))$ are at worst the poles of $(df)_{\mathfrak{P},i}$. This suffices except in the case $\mathfrak{P}=\mathfrak{Q}$, $i=j$ and $p\not|v_C(f)$, but in this case we can do better. By 12.36 this can happen only for $\mathbf{j}=0$ and the poles along C of $(df)_{\mathfrak{P},i}$ are the poles of $(d\delta)_{\mathfrak{P},i}/\delta^{n+1}$ see 12.31.1. The fact that $v_C\tilde{\sigma}^{[0]}_{\mathfrak{P},i}\left((d\delta)_{\mathfrak{P},i}/\delta^{n+1}\right)=v_C(f)-(p^{f_{\mathfrak{P}}}-1)$ follows from (12.32.1), (12.33.1) and 12.34.

By 9.3 the modular form $a\left(\chi^2_{\mathfrak{P},i}\right)$ defined in 7.8 extends to a section of $\overline{\phi}^*\left(\mathcal{L}_{\chi^2_{\mathfrak{P},i}}\right)$ over $\mathfrak{M}^*$, which we denote in the same way. It is non-vanishing on $\mathfrak{M}$ and locally on $\mathfrak{M}^*\backslash\mathfrak{M}$ it satisfies

$$a\left(\chi_{\mathfrak{P},i}\right)^{p^{f_{\mathfrak{P}}}-1}=h_{\mathfrak{P},i+1}^{p^{f_{\mathfrak{P}}-1}}h_{\mathfrak{P},i+2}^{p^{f_{\mathfrak{P}}-2}}\cdots h_{\mathfrak{P},i}.$$

Hence, $v_C\left(a(\chi^2_{\mathfrak{P},i})\right)$ is equal to 0 if $\mathfrak{P}\neq\mathfrak{Q}$, it is equal to $2p^{f_{\mathfrak{P}}-r}$ if $\mathfrak{P}=\mathfrak{Q}$ and $i\neq j$, and it is equal to 2, if $\mathfrak{P}=\mathfrak{Q}$ and $i=j$. We conclude using the previous Proposition.

12.38 Definition. *(The operator $\Theta^{[\mathrm{j}]}_{\mathfrak{P},i}$ on modular forms).* Let $\mathfrak{P}$ be a prime over p, let $1\leq i\leq f_{\mathfrak{P}}$ and let $0\leq\mathbf{j}\leq e_{\mathfrak{P}}-1$. Define the subspace of $\mathfrak{I}$-polarized modular forms of weight ψ

$$\mathbf{M}\left(k,\mu_N,\psi\right)^{\mathfrak{P},[\mathrm{j}]}:=\left\{f\in\mathbf{M}\left(k,\mu_N,\psi\right)|r(f)\in\Gamma\left(\mathfrak{M},O^{\mathfrak{P},[\mathrm{j}]}_{\mathfrak{M}}\right)\right\},$$

where $r(f)$ is the regular function on $\mathfrak{M}$ defined in 7 19 and $\Gamma\left(\mathfrak{M},O^{\mathfrak{P},[\mathrm{j}]}_{\mathfrak{M}}\right)$ is defined in 12.12. For $f\in\mathbf{M}\left(k,\mu_N,\psi\right)^{\mathfrak{P},[\mathrm{j}]}$ define

$$\Theta^{[\mathrm{j}]}_{\mathfrak{P},i}(f):=\Theta^{[\mathrm{j}]}_{\mathfrak{P},i}\left(r(f)\right)\cdot a(\psi)\cdot a\left(\chi^2_{\mathfrak{P},i}\right)\cdot h_{\mathfrak{P},i}.$$

We have

$$\Theta^{[\mathrm{j}]}_{\mathfrak{P},i}(f)\in\Gamma\Big(\mathfrak{M},\mathcal{L}_{\psi\chi^p_{\mathfrak{P},i-1}\chi_{\mathfrak{P},i}}\Big).$$

12.39 Theorem. *The notation is as in 12.38.*

1. *The section $\Theta^{[\mathrm{j}]}_{\mathfrak{P},i}(f)$ over $\mathfrak{M}(1,1)^{\mathrm{Kum}}$ descends to a section of the line bundle $\mathcal{L}_{\psi\chi^p_{\mathfrak{P},i-1}\chi_{\mathfrak{P},i}}$ over $\mathfrak{M}\left(k,\mu_N\right)^{\mathrm{ord}}$;*
2. *the section $\Theta^{[\mathrm{j}]}_{\mathfrak{P},i}(f)$ extends to a section of $\mathcal{L}_{\psi\chi^p_{\mathfrak{P},i-1}\chi_{\mathfrak{P},i}}$ over $\mathfrak{M}(k,\mu_N)^{\mathrm{R}}$.*

Hence, we obtain a k-derivation:

$$\Theta^{[j]}_{\mathfrak{P},i}: \bigoplus_{\psi} \mathbf{M}\big(k,\mu_N,\psi\big)^{\mathfrak{P},[j]} \longrightarrow \bigoplus_{\psi} \mathbf{M}\big(k,\mu_N,\psi\chi^p_{\mathfrak{P},i-1}\chi_{\mathfrak{P},i}\big).$$

Proof: Part (1) follows from the description of the action of the Galois group G on functions and on modular forms given in 7.8. Part (2) follows from 12.37.

12.40 Corollary. *The notation is as in 12.38 and in 12.16. Let $f \in \mathbf{M}\big(k,\mu_N,\psi\big)$ be a $\mathfrak{J}$-polarized modular form of level N and weight ψ. Suppose that the q-expansion of f at a $\mathfrak{J}$-polarized unramified cusp $\big(\mathfrak{A},\mathfrak{B},\varepsilon,\mathrm{j}_\varepsilon\big)$ is*

$$f\big(\mathbf{Tate}(\mathfrak{A},\mathfrak{B}),\varepsilon,\mathrm{j}_\varepsilon\big) = \sum_\nu a_\nu q^\nu.$$

Then

1. $f \in \mathbf{M}\big(k,\mu_N,\psi\big)^{\mathfrak{P},[j]}$ *if and only if* $a_\nu = 0$ *for all* $\nu \notin \mathfrak{P}^{\mathrm{j}} M$;
2. *if* $f \in \mathbf{M}\big(k,\mu_N,\psi\big)^{\mathfrak{P},[j]}$, *the q-expansion of $\Theta^{[j]}_{\mathfrak{P},i}(f)$ at the same cusp is*

$$\Theta^{[j]}_{\mathfrak{P},i}(f)\big(\mathbf{Tate}(\mathfrak{A},\mathfrak{B}),\varepsilon,\mathrm{j}_\varepsilon\big) = \sum_\nu \tilde{\chi}^{[j]}_{\mathfrak{P},i}(\nu) a_\nu q^\nu.$$

See 12.21 for the definition of $\tilde{\chi}^{[j]}_{\mathfrak{P},i}$.

Proof: It follows from the definition of $\Theta_{\mathfrak{P},i}$ given in 12.38 and 12.21.

12.41 Corollary. *The theta operators $\Theta^{[j]}_{\mathfrak{P},i}$ commute for different primes $\mathfrak{P}$ and different $1 \le i \le f_{\mathfrak{P}}$.*

12.42 *Comparison with Katz's definition.* Katz's definition of theta operators extends to Hilbert modular form in characteristic p; see [Ka4, §2.6]. Note however that our operators $\Theta_{\mathfrak{P},i}$ in characteristic p and Katz's theta operators change the weights in a different way; compare 12.38 with [Ka4, Cor. 2.6.25]. Indeed, our theta operators in characteristic p are Katz's operators multiplied by suitable partial Hasse invariants in order to ensure that they send holomorphic modular forms to holomorphic modular forms. Instead, Katz is interested only in modular forms defined over the *ordinary locus* where the holomorphicity is not an issue.

13 The operator V.

In this section we suppose that $k \subset \bar{\mathbf{F}}_p$.

13.1 *The definition.* Let $g \in \mathbf{M}\big(k,\mu_N,\psi\big)$ be a $\mathfrak{J}$-polarized modular form over k of level μ_N and of weight ψ. Let R be a k-algebra. Let

$$\underline{A} := \big(A,\iota,\lambda,\varepsilon\big), \qquad \omega \in \mathrm{H}^0\big(R,\Omega^1_{A/R}\big),$$

be a $\mathfrak{J}$-polarized abelian scheme over R with O_L-action and μ_N-level structure, as defined in 3.2, and a generator ω of $\mathrm{H}^0\big(R,\Omega^1_{A/R}\big)$ as a free $O_L \otimes_{\mathbf{Z}} R$-module. Define

$$V(g)\big(\underline{A},\omega\big) := g\big(\underline{A}^{(p)},\omega^{(p)}\big),$$

where the superscript (p) stands for the base change by the *absolute Frobenius* on R.

13.2 Definition. *The notation is as in 2.2. Let* $\mathrm{F}^{\mathrm{abs}}\colon \mathcal{G}_{\mathbf{F}_p} \to \mathcal{G}_{\mathbf{F}_p}$ *be the absolute Frobenius morphism on* $\mathcal{G}_{\mathbf{F}_p}$. *Let* $\mathrm{F}_k^{\mathrm{abs}} := \mathrm{F}^{\mathrm{abs}} \times_{\mathrm{Spec}(\mathbf{F}_p)} \mathrm{Spec}(k)$ *be the base change of* $\mathrm{F}^{\mathrm{abs}}$ *to* $\mathrm{Spec}(k)$. *Define the following endomorphism of the group of characters* $\mathbb{X}_k$ *by*

$$\left(\chi\colon \mathcal{G}_k \to \mathbf{G}_{m,k}\right) \longmapsto \left(\chi^{(p)} := \chi \circ \mathrm{F}_k^{\mathrm{abs}}\right).$$

It preserves the universal characters defined in 4.1.

13.3 Remark. Suppose that $k = \bar{\mathbf{F}}_p$. Then the group of characters $\mathbb{X}_k$ of $\mathcal{G}_k$ is endowed with a $\mathrm{Gal}(k/\mathbf{F}_p)$-action. In particular, the absolute Frobenius $\sigma \in \mathrm{Gal}(k/\mathbf{F}_p)$ defines an action on $\mathbb{X}_k$. Explicitly:

$$\sigma^*\colon \chi \mapsto \sigma^{-1} \circ \chi \circ \sigma.$$

Such action preserves the universal characters since

$$\sigma^*\left(\chi_{\mathfrak{P},i}\right) = \chi_{\mathfrak{P},i-1}.$$

Indeed, if $a \otimes 1 \in O_L \otimes_{\mathbf{Z}} \bar{\mathbf{F}}_p$, then

$$\left(\sigma^{-1} \circ \chi_{\mathfrak{P},i} \circ \sigma\right)(a \otimes 1) = \left(\sigma^{-1} \circ \chi_{\mathfrak{P},i}\right)(a \otimes 1) = \chi_{\mathfrak{P},i-1}(a).$$

For any $\chi \in \mathbb{X}_k$, we have

$$\chi^{(p)} = \left(\sigma^*(\chi)\right)^p.$$

13.4 *Examples.* For p inert, $\mathbb{X}_k$ is freely generated by $\{\chi_1, \dots, \chi_g\}$ and $\chi_i^{(p)} = \chi_{i-1}^p$, where $\chi_1^{(p)} = \chi_g^p$. If p is totally ramified, then $\mathbb{X}_k$ is freely generated by one character Ψ and $\Psi^{(p)} = \Psi^p$.

13.5 Proposition. *Let* $g \in \mathbf{M}(k, \mu_N, \psi)$, *then* $V(g) \in \mathbf{M}(k, \mu_N, \psi^{(p)})$.

Proof: We need to verify properties (I)-(III) of 5.1. The first two clearly hold. For (III) consider any $\underline{A}$ and any ω as in 13.1 and any $\gamma \in \mathcal{G}_k(R) = \left(O_L \otimes_{\mathbf{Z}} R\right)^*$. Then, with the notation $\gamma^{(p)} := \mathrm{F}_k^{\mathrm{abs}}(\gamma)$, we have

$$\begin{aligned} V(g)\left(\underline{A}, \gamma^{-1}\omega\right) &= g\left(\underline{A}^{(p)}, (\gamma^{(p)})^{-1}\omega^{(p)}\right) \\ &= \psi\left(\gamma^{(p)}\right) g\left(\underline{A}^{(p)}, \omega^{(p)}\right) \\ &= \psi^{(p)}(\gamma)\left(V(g)\left(\underline{A}, \omega\right)\right). \end{aligned}$$

We want to compute the effect of V on q-expansions. To do this we introduce an auxiliary operator.

13.6 Definition. *Let* σ *be the absolute Frobenius on* k. *Consider the induced self-equivalence on the category of schemes over* k *given by*

$$S \longmapsto S^\sigma := S \underset{k,\sigma}{\times} k.$$

Let g be a $\mathfrak{J}$-polarized modular form of weight ψ and level μ_N over k. Define the rule $\sigma^*(g)$ by

$$\sigma^*(g)(\underline{A},\omega) := \Big(g(\underline{A}^\sigma,\omega^\sigma)\Big)^{\sigma^{-1}},$$

where $\underline{A}$ and ω are as in 13.1.

13.7 Remark. Let $\underline{A}$ be defined over a k-algebra R. We have defined two different objects $A^{(p)} \to \mathrm{Spec}(R)$ and $A^\sigma \to \mathrm{Spec}(R^\sigma)$. Letting σ be the Frobenius on R, they are defined by the following cartesian diagrams

$$\begin{array}{ccc} A^{(p)} & \longrightarrow & A \\ \downarrow & & \downarrow \\ \mathrm{Spec}(R) & \xrightarrow{\sigma} & \mathrm{Spec}(R), \end{array} \qquad \begin{array}{ccc} A^{\sigma} & \longrightarrow & A \\ \downarrow & & \downarrow \\ \mathrm{Spec}(k) & \xrightarrow{\sigma} & \mathrm{Spec}(k). \end{array}$$

13.8 Lemma. *The rule $\sigma^*(g)$ defines a $\mathfrak{J}$-polarized modular form of weight $\sigma^*(\psi)$ as defined in 13.3. Moreover, if the q-expansion of g at a $\mathfrak{J}$-polarized $\mathbf{F}_p$-rational unramified cusp $(\mathfrak{A},\mathfrak{B},\varepsilon,\mathrm{j})$ is*

$$g\big(\mathbf{Tate}(\mathfrak{A},\mathfrak{B}),\varepsilon,\mathrm{j}\big) = \sum_\nu a_\nu q^\nu,$$

then the q-expansion of $\sigma^*(g)$ at the same cusp is

$$\sigma^*(g)\big(\mathbf{Tate}(\mathfrak{A},\mathfrak{B}),\varepsilon,\mathrm{j}\big) = \sum_\nu a_\nu^{\sigma^{-1}}\, q^\nu.$$

Proof: To prove that $\sigma^*(g)$ defines a modular form, one has to check that (I)–(III) of 5.1 hold. Properties (I) and (II) are clearly satisfied. We verify that (III) holds as well. For any $\mathfrak{J}$-polarized Hilbert-Blumenthal abelian scheme $\underline{A}$ over R with μ_N-level structure and for any generator ω of the relative differentials and for any $\gamma \in \big(O_L \otimes_{\mathbf{Z}} R\big)^*$, we have:

$$\begin{aligned} \sigma^*(g)(\underline{A},\gamma^{-1}\omega) &= \Big(g\big(\underline{A}^\sigma,(\gamma^\sigma)^{-1}\omega^\sigma\big)\Big)^{\sigma^{-1}} \\ &= \Big(\psi(\gamma^\sigma)\Big)^{\sigma^{-1}}\Big(g(\underline{A}^\sigma,\omega^\sigma)\Big)^{\sigma^{-1}} \\ &= \Big(\sigma^*(\psi)(\gamma)\Big)\Big(\sigma^*(g)(\underline{A},\omega)\Big). \end{aligned}$$

For the assertion on q-expansions note that

$$\begin{aligned} \sigma^*(g)\big(\mathbf{Tate}(\mathfrak{A},\mathfrak{B}),\varepsilon,\mathrm{j}\big) &= \Big(g\big(\mathbf{Tate}(\mathfrak{A},\mathfrak{B})^\sigma,\varepsilon^\sigma,(\mathrm{j})^\sigma\big)\Big)^{\sigma^{-1}} \\ &= \Big(g\big(\mathbf{Tate}(\mathfrak{A},\mathfrak{B}),\varepsilon,\mathrm{j}\big)\Big)^{\sigma^{-1}}, \end{aligned}$$

since Tate objects are defined over $\mathbf{F}_p$.

13.9 Proposition. *The notation is as in 13.1. Fix a $\mathfrak{J}$-polarized unramified $\mathbf{F}_p$-rational cusp $(\mathfrak{A},\mathfrak{B},\varepsilon,\mathrm{j})$. Suppose that g has q-expansion*

$$g\big(\mathbf{Tate}(\mathfrak{A},\mathfrak{B}),\varepsilon,\mathrm{j}\big) = \sum_\nu a_\nu q^\nu,$$

then the q-expansion of $V(g)$ at the same cusp is

$$V(g)\big(\mathbf{Tate}(\mathfrak{A},\mathfrak{B}),\varepsilon,\mathrm{j}\big)=\sum_{\nu} a_\nu\, q^{p\nu}.$$

Proof: We use the notations of 6.3. Let $k\big((\mathfrak{A},\mathfrak{B},\sigma_\beta)\big):=\mathbf{Z}\big((\mathfrak{A},\mathfrak{B},\sigma_\beta)\big)\otimes_{\mathbf{Z}} k$. We can factor the absolute Frobenius $\mathrm{F}^{\mathrm{abs}}$ on $k\big((\mathfrak{A},\mathfrak{B},\sigma_\beta)\big)$ as follows:

$$\begin{array}{ccccc} k[\![q^\nu]\!] & \xrightarrow{\sigma} & k[\![q^\nu]\!] & \xrightarrow{\xi} & k[\![q^\nu]\!] \\ \sum_\nu a_\nu q^\nu & \mapsto & \sum_\nu a_\nu^p q^\nu & & \\ & & \sum_\nu b_\nu q^\nu & \mapsto & \sum_\nu b_\nu q^{p\nu}. \end{array}$$

Note that ξ is a homomorphism of k-algebras. In particular,

$$\begin{aligned} V(g)\big(\mathbf{Tate}(\mathfrak{A},\mathfrak{B}),\varepsilon,\mathrm{j}\big) &= g\big(\mathbf{Tate}(\mathfrak{A},\mathfrak{B})^{(p)},\varepsilon^{(p)},(\mathrm{j})^{(p)}\big)\\ &= g\left(\big((\mathbf{Tate}(\mathfrak{A},\mathfrak{B}),\varepsilon)^\sigma\big)^\xi,\big((\mathrm{j})^\sigma\big)^\xi\right)\\ &= \xi\left(g\big(\mathbf{Tate}(\mathfrak{A},\mathfrak{B})^\sigma,\varepsilon^\sigma,(\mathrm{j})^\sigma\big)\right) && \text{(by 5.1}(II))\\ &= \xi\circ\sigma\left(\sigma^*(g)\big(\mathbf{Tate}(\mathfrak{A},\mathfrak{B}),\varepsilon,\mathrm{j}\big)\right) && \text{(by 13.6)}\\ &= \xi\circ\sigma\left(\sum_\nu a_\nu^{\sigma^{-1}}\, q^\nu\right) && \text{(by 13.8)}\\ &= \sum_\nu a_\nu\, q^{p\nu}. \end{aligned}$$

13.10 *Katz's V operator.* In [Ka4, §1.11.21] one finds a more general notion of Frobenius operator on Katz's $\mathfrak{I}$-polarized p-adic Hilbert modular forms in the sense of 11.4

$$V\colon \mathbf{M}(R,\mu_N,\chi)^{p-\mathrm{adic}}\longrightarrow \mathbf{M}(R,\mu_N,\chi)^{p-\mathrm{adic}},$$

where R is as in 10.1. It is defined as follows.

For $n,m\in\mathbf{N}$ let

$$\big(\mathrm{A}^{\mathrm{U}},\iota^{\mathrm{U}},\lambda^{\mathrm{U}},\varepsilon^{\mathrm{U}}_{Np^n}\big)\longrightarrow \mathfrak{M}(m,n)$$

be the universal $\mathfrak{I}$-polarized abelian scheme with real multiplication by O_L and μ_{Np^n}-level structure. Define for $n\geq 1$

$$\pi\colon \mathrm{A}^{\mathrm{U}}\longrightarrow (\mathrm{A}^{\mathrm{U}})':=\mathrm{A}^{\mathrm{U}}/\big(\mathrm{D}_L^{-1}\underset{\mathbf{Z}}{\otimes}\mu_p\big).$$

Then, $(\mathrm{A}^{\mathrm{U}})'$ inherits a canonical O_L-action $(\iota^{\mathrm{U}})'$ and a canonical $\mu_N\times\mu_{p^{n-1}}$-level structure $(\varepsilon^{\mathrm{U}}_{Np^n})'$. By [Ka4, Lem. 1.11.6] it inherits a canonical polarization data λ'. For $m=1$ we have that

$$\big((\mathrm{A}^{\mathrm{U}})',\ldots,(\varepsilon^{\mathrm{U}}_{Np^n})'\big)\cong\big((\mathrm{A}^{\mathrm{U}})^{(p)},\ldots,(\varepsilon^{\mathrm{U}}_{Np^{n-1}})^{(p)}\big)\underset{\mathfrak{M}(1,n-1)}{\times}\mathfrak{M}(1,n).$$

Futhermore, for any $n\geq 1$ there exists a unique morphism of schemes over $R/\mathfrak{m}^m$

$$F_n\colon \mathfrak{M}(m,n)\longrightarrow \mathfrak{M}(m,n-1)$$

such that

$$\big((\mathrm{A}^{\mathrm{U}})',(\iota^{\mathrm{U}})',(\lambda^{\mathrm{U}})',(\varepsilon^{\mathrm{U}}_{Np^n})'\big)\cong\left(\mathrm{A}^{\mathrm{U}},\iota^{\mathrm{U}},\lambda^{\mathrm{U}},\varepsilon^{\mathrm{U}}_{Np^{n-1}}\right)\underset{\mathfrak{M}(m,n-1),\mathrm{F}_n}{\times}\mathfrak{M}(m,n).$$

With the notation of 11.1, let $\alpha \in \Gamma_n$. By construction the diagram

$$\begin{array}{ccc} \mathfrak{M}(m,n) & \xrightarrow{\alpha} & \mathfrak{M}(m,n) \\ {\scriptstyle F_n}\downarrow & & \downarrow{\scriptstyle F_n} \\ \mathfrak{M}(m,n-1) & \xrightarrow{\alpha} & \mathfrak{M}(m,n-1) \end{array}$$

is commutative. This gives a well defined Γ-equivariant morphism

$$V\colon O_{\mathfrak{M}(m,\infty)} \longrightarrow O_{\mathfrak{M}(m,\infty)}, \qquad f \mapsto f \circ F,$$

where $F := \lim F_n$. Moreover, the following diagram commutes

$$\begin{array}{ccc} O_{\mathfrak{M}(m+1,\infty)} & \xrightarrow{V} & O_{\mathfrak{M}(m+1,\infty)} \\ \downarrow & & \downarrow \\ O_{\mathfrak{M}(m,\infty)} & \xrightarrow{V} & O_{\mathfrak{M}(m,\infty)}. \end{array}$$

Let $f = \{f_m\}_n$ be a $\mathfrak{I}$-polarized p-adic Hilbert modular form of level μ_N and weight χ over R, with $f_m \in \Gamma(\mathfrak{M}(m,\infty), O_{\mathfrak{M}(m,\infty)})$ transforming via χ under the action of Γ. It follows that the sequence $\{V(f_m)\}_{m\in\mathbf{N}}$ defines a p-adic modular form $V(f)$ of the *same* weight χ, $V(f) \in \mathbf{M}(R,\mu_N,\chi)^{p-\mathrm{adic}}$.

Fix a $\mathfrak{I}$-polarized unramified cusp $(\mathfrak{A},\mathfrak{B},\varepsilon_{p^\infty N},\mathrm{j}_\varepsilon)$. By [Ka4, §1.11.23] the operator V changes the q-expansions at this cusp according to the rule

$$\sum_\nu a_\nu q^\nu \longmapsto \sum_\nu a_\nu q^{p\nu}.$$

13.11 Remark. Let f be a Serre p-adic modular form over F of level μ_N; see 10.8. Then, $\Theta_{\mathfrak{P},i}^{[j]}(f)$ and $V(f)$ are Serre p-adic modular forms over F if f is either a cusp form, or of weight $\chi \in \mathbb{X}$, or of weight $\mathbf{Nm}^z$ with $z \in \mathbf{Z}_p$. This follows from 11.11.

13.12 Lemma. *Let $f \in \mathbf{M}(k,\mu_N,\chi)$. We have*

$$r\big(V(f)\big) = V\big(r(f)\big),$$

where the V on the left hand side is the one defined in 13.1, while the one defined on the right hand side is Katz's V operator on regular functions on $\mathfrak{M}(1,\infty)$ and

$$r\colon \bigoplus_\psi \mathbf{M}(k,\mu_N,\psi) \longrightarrow \Gamma\big(\mathfrak{M}(1,1), O_{\mathfrak{M}(1,1)}\big)$$

is the map $\sum g_\psi \to \sum_\psi g_\psi/a(\psi)$ defined in 7.19.

Proof: By 7.9 the q-expansion of the $a(\psi)$'s can be assumed to be 1. The lemma follows by comparing the effect on q-expansions of Katz's V operator and our V operator; see 13.9.

13.13 Remark. If $f \in \mathbf{M}(k,\mu_N,\chi)$, then $V(f) \in \mathbf{M}(k,\mu_N,\chi^{(p)})$. Note that $\chi^{(p)}\chi^{-1} \in \mathbb{X}_k(1)$ in the notation of 4.11. In particular, $r\big(V(f)\big)$ and $V\big(r(f)\big)$ have the same weight as functions on $\mathfrak{M}(1,1)$. This justifies our approach to the operator V; it allows to control how the weight changes (see 13.5).

13.14 Remark. Recall that $\mathfrak{M}(k,\mu_N)$ is defined over $\mathbf{F}_p$. Let $\mathrm{F}^{\mathrm{abs}}$ be the absolute Frobenius morphism on $\mathfrak{M}(\mathbf{F}_p,\mu_N)$ and let $\mathrm{F}_k^{\mathrm{abs}}$ be its base-change to k. Using the moduli property of $\mathfrak{M}(k, \mu_N)$, one sees that if g is a modular form of level μ_N and weight $\mathbb{1}$ over k i. e., $g \in \Gamma(\mathfrak{M}(k, \mu_N), O_{\mathfrak{M}(k,\mu_N)})$, then $V(g) = \mathrm{F}_k^{\mathrm{abs},*}(g)$.

Analogously, let $\mathrm{F}^{\mathrm{abs}}$ be the absolute Frobenius morphism on $\mathfrak{M}(\mathbf{F}_p, \mu_{pN})$ and let $\mathrm{F}_k^{\mathrm{abs}}$ be its base-change to k. It follows from 13.10 that $\mathrm{F}_k^{\mathrm{abs},*}: O_{\mathfrak{M}(1,1)} \to O_{\mathfrak{M}(1,1)}$ extends the operator $V: O_{\mathfrak{M}(1,0)} \to O_{\mathfrak{M}(1,1)}$. Note that this is in agreement with 13.12.

14 The operator U.

We use previously introduced notation

$$\mathfrak{M} := \mathfrak{M}(1,1)^{\mathrm{Kum}}, \qquad \mathfrak{M}' := \mathfrak{M}(1,0).$$

Then $\phi: \mathfrak{M} \to \mathfrak{M}'$ is a Galois cover with group

$$G \cong \prod_{\mathfrak{P}|(p)} (O_L/\mathfrak{P})^*.$$

14.1 Definition. *Let $\mathfrak{P}$ be a prime of O_L over p. Let $0 \leq \mathfrak{j} \leq e_{\mathfrak{P}} - 1$ be an integer. Define an operator*

$$\Lambda(\mathfrak{P}, \mathfrak{j}): \Gamma(\mathfrak{M}, O_{\mathfrak{M}}^{\mathfrak{P},[\mathfrak{j}]}) \longrightarrow \Gamma(\mathfrak{M}, O_{\mathfrak{M}}^{\mathfrak{P},[\mathfrak{j}+1]})$$

(see 12.12 for the notation, extended by the same formula for $\mathfrak{j}+1 = e_{\mathfrak{P}}$) as follows. Choose $\psi = \prod_{\mathfrak{P},i} \chi_{\mathfrak{P},i}^{a_{\mathfrak{P},i}} \in \mathbb{X}_k(1)$, in the notation of 4.11, with $a_{\mathfrak{P},i} \geq 0$. Let

$$\Lambda(\mathfrak{P}, \mathfrak{j}) := \mathrm{Id} - \prod_{i=1}^{f_{\mathfrak{P}}} (\Theta_{\mathfrak{P},i}^{[\mathfrak{j}]})^{a_{\mathfrak{P},i}}.$$

14.2 Lemma. *The operator $\Lambda(\mathfrak{P}, \mathfrak{j})$ is well defined.*

1. *It has the following effect on q-expansions. If $f = a_0 + \sum_{\nu \in (\mathfrak{P}^{\mathfrak{j}}\mathfrak{A}\mathfrak{B})^+} a_\nu q^\nu$ at the $\mathfrak{I}$-polarized unramified cusp $(\mathfrak{A}, \mathfrak{B}, \varepsilon_{pN}, \mathrm{j}_\varepsilon)$, then*

$$\Lambda(\mathfrak{P}, \mathfrak{j})(f) = \sum_{\nu \in (\mathfrak{P}^{\mathfrak{j}+1}\mathfrak{A}\mathfrak{B})^+} a_\nu q^\nu$$

at this cusp;

2. *let $\chi: G \to k^*$ be a character and let f be an eigenfunction for G with character χ. Then $\Lambda(\mathfrak{P}, \mathfrak{j})(f)$ is also an eigenfunction for G with character χ.*

Proof: Choose ψ as in 14.1. We calculate the effect of $\Lambda(\mathfrak{P}, \mathfrak{j})$ on q-expansions using that the operators $\Theta_{\mathfrak{P},i}$, for different i's, commute by 12.41. Suppose $f \in \Gamma(\mathfrak{M}, O_{\mathfrak{M}}^{\mathfrak{P},[\mathfrak{j}]})$ i. e., by 12.21, that $a_\nu = 0$ if $\nu \notin \mathfrak{P}^{\mathfrak{j}}\mathfrak{A}\mathfrak{B}$. It follows from 12.21 that the q-expansion of

$$\left(\mathrm{Id} - \prod_{i=1}^{f_{\mathfrak{P}}} (\Theta_{\mathfrak{P},i}^{[\mathfrak{j}]})^{a_{\mathfrak{P},i}}\right)(f)$$

at the given cusp is

$$\sum_{\nu}\Big(1-\big(\prod_{i=1}^{f_{\mathfrak{P}}}(\tilde{\chi}_{\mathfrak{P},i}^{[\mathfrak{j}]})^{a_{\mathfrak{P},i}}\big)(\nu)\Big)a_\nu q^\nu = \sum_{\nu\in(\mathfrak{P}^{\mathfrak{j}+1}\mathfrak{A}\mathfrak{B})^+} a_\nu q^\nu .$$

The last equality holds since

$$\prod_{i=1}^{f_{\mathfrak{P}}}\big(\tilde{\chi}_{\mathfrak{P},i}^{[\mathfrak{j}]}(\nu)\big)^{a_{\mathfrak{P},i}} = \begin{cases} 0 & \text{if } \nu\in\mathfrak{P}^{\mathfrak{j}+1}\mathfrak{A}\mathfrak{B},\\ 1 & \text{otherwise.}\end{cases}$$

In particular, by 12.21,

$$\Lambda(\mathfrak{P},\mathfrak{j})(f)\in\Gamma\left(\mathfrak{M}, O_{\mathfrak{M}}^{\mathfrak{P},[\mathfrak{j}+1]}\right)$$

and the definition of $\Lambda(\mathfrak{P},\mathfrak{j})$ does not depend on the choice of ψ. Part (2) follows from 12.14.

14.3 Definition. *Define an operator*

$$\Lambda\colon\Gamma(\mathfrak{M},O_{\mathfrak{M}})\longrightarrow\Gamma(\mathfrak{M},O_{\mathfrak{M}})$$

by

$$\Lambda := \circ_{\mathfrak{P}|p}\Big(\Lambda(\mathfrak{P},e_{\mathfrak{P}}-1)\circ\cdots\circ\Lambda(\mathfrak{P},0)\Big).$$

14.4 Proposition. *The operator Λ is well defined.*

1. *It has the following effect on q-expansions. If $f=\sum_\nu a_\nu q^\nu$ at the $\mathfrak{J}$-polarized unramified cusp $(\mathfrak{A},\mathfrak{B},\varepsilon_{pN},\mathfrak{j}_\varepsilon)$, then*

   $$\Lambda(f)=\sum_\nu a_{p\nu}q^{p\nu}$$

   at this cusp;
2. *let $\chi\colon G\to k^*$ be a character and let f be an eigenfunction for G with character χ. Then $\Lambda(f)$ is also an eigenfunction for G with character χ.*

Proof: This follows from 14.2.

14.5 Proposition. *Let f be a regular function on $\mathfrak{M}$. There exists a unique regular function a on $\mathfrak{M}$ such that*

$$V(a)=\Lambda(f).$$

Moreover, if f is an eigenfunction for G with character χ, then a is also an eigenfunction for G with character χ.

Proof: Let K be the function field of $\mathfrak{M}$. We have a commutative diagram:

$$\begin{array}{ccc} K & \xrightarrow{d} & \Omega^1_{K/k}\\ \downarrow & & \downarrow\\ k\big((\mathfrak{A},\mathfrak{B},\sigma_\beta)\big) & \xrightarrow{d} & \Omega^1_{k\left((\mathfrak{A},\mathfrak{B},\sigma_\beta)\right)/k}\end{array}.$$

The left vertical arrow is injective being a morphism of fields. By [Ra, Thm. 5.1] the schemes $S^{\wedge}_{\sigma_\beta}$, defined in 6.3, for a suitable choice of the cone decomposition $\{\sigma_\beta\}_\beta$ appear as formal completions of boundary components of smooth toroidal compactifications of $\mathfrak{M}$. In particular,

$$\Omega^1_{K/k} \hookrightarrow \Omega^1_{k((\mathfrak{A},\mathfrak{B},\sigma_\beta))/k}$$

is injective. It now follows from 14.4 that

$$\Omega^1_{K/k} \ni d\left(\Lambda(f)\right) = 0.$$

We conclude that

$$\Lambda(f) \in \Gamma\big(\mathfrak{M}, O_{\mathfrak{M}}\big) \cap K^p k;$$

see [La, Ch. X, Prop. 7.4]. Since $\mathfrak{M}$ is normal and affine

$$\Gamma\big(\mathfrak{M}, O_{\mathfrak{M}}\big) \cap K^p k := \Gamma\big(\mathfrak{M}, O_{\mathfrak{M}}\big)^p k.$$

We conclude by 13.14.

14.6 Definition. *Let f be a regular function on $\mathfrak{M}$. Define $U(f)$ to be the unique regular function on $\mathfrak{M}$ such that*

$$V\Big(U(f)\Big) = \Lambda(f).$$

Its existence and uniqueness is guaranteed by 14.5.

14.7 Theorem. *Let $\mathbf{M}(k,\mu_N,\chi)^{\mathrm{ord}}$ be the space of $\mathfrak{I}$-polarized modular forms over k of weight χ defined on $\mathfrak{M}(k,\mu_N)^{\mathrm{ord}}$. There exists a (unique) k-linear operator*

$$U\colon \mathbf{M}(k,\mu_N,\chi)^{\mathrm{ord}} \longrightarrow \mathbf{M}(k,\mu_N,\chi)^{\mathrm{ord}}$$

with the following effect on q-expansions. Let $f \in \mathbf{M}(k,\mu_N,\chi)^{\mathrm{ord}}$. Suppose that its q-expansion at a $\mathfrak{I}$-polarized unramified cusp $(\mathfrak{A},\mathfrak{B},\varepsilon_{pN},\mathrm{j}_\varepsilon)$ is

$$f\big(\mathbf{Tate}(\mathfrak{A},\mathfrak{B}),\varepsilon_{pN},\mathrm{j}_\varepsilon\big) = \sum_\nu a_\nu q^\nu.$$

Then the q-expansion of $U(f)$ at the same cusp is

$$U(f)\big(\mathbf{Tate}(\mathfrak{A},\mathfrak{B}),\varepsilon_{pN},\mathrm{j}_\varepsilon\big) = \sum_\nu a_{p\nu}\, q^\nu.$$

Moreover, if $\chi = \prod_{\mathfrak{P},i} \chi_{\mathfrak{P},i}^{a_{\mathfrak{P},i}}$ and $a_{\mathfrak{P},i} \geq (e_{\mathfrak{P}}-1)p + e_{\mathfrak{P}} + 1$ for all primes $\mathfrak{P}$ and all integers $1 \leq i \leq f_{\mathfrak{P}}$, then

$$f \in \mathbf{M}(k,\mu_N,\chi) \quad \Longrightarrow \quad U(f) \in \mathbf{M}(k,\mu_N,\chi).$$

Proof: Let U be the operator on functions on $\mathfrak{M}$ defined in 14.6. Let

$$U(f) := U\big(r(f)\big)a(\chi),$$

where $r(f)$ is the regular function on $\mathfrak{M}$ associated to f as in 7.19 and $a(\chi)$ is the modular form on $\mathfrak{M}$ defined in 7.8. In particular, $U(f)$ is a modular form on $\mathfrak{M}$. By 7.8 it descends to a modular form on the ordinary locus $\mathfrak{M}' = \mathfrak{M}(k,\mu_N)^{\mathrm{ord}}$

of $\mathfrak{M}(k,\mu_N)$. It has weight χ. By 14.4, 14.6 and 13.9, we conclude that the effect on q-expansions is as claimed in the theorem.

By comparing weights and q-expansions we conclude the following equality of modular forms

$$h^{p+1}V(U(f)) = h_{\chi^*}\cdot\left(\prod_{\mathfrak{P}}\prod_{\mathrm{j}=0}^{e_{\mathfrak{P}}-1}\left(\prod_{i=1}^{f_{\mathfrak{P}}}h_{\mathfrak{P},i}^{p+1}-\prod_{i=1}^{f_{\mathfrak{P}}}(\Theta_{\mathfrak{P},i}^{[\mathrm{j}]})^{p-1}\right)\right)(f), \qquad (14.7.1)$$

where $\chi^*=\prod_{\mathfrak{P},i}\psi_{\mathfrak{P},i}^{a_{\mathfrak{P},i}}$, $h_{\mathfrak{P},i}$ are the partial Hasse invariants and h_{χ^*} and h are defined as in 7.12.

Assume that $f\in\mathbf{M}(k,\mu_N,\chi)$ with $\chi=\prod_{\mathfrak{P},i}\chi_{\mathfrak{P},i}^{a_{\mathfrak{P},i}}$ and $a_{\mathfrak{P},i}\geq(e_{\mathfrak{P}}-1)p+e_{\mathfrak{P}}+1$ for all primes $\mathfrak{P}$ and all integers $1\leq i\leq f_{\mathfrak{P}}$. Consider the equality of meromorphic modular forms:

$$\prod_{\mathfrak{P}}\left(\prod_{i=1}^{f_{\mathfrak{P}}}h_{\mathfrak{P},i}^{e_{\mathfrak{P}}(p+1)-a_{\mathfrak{P},i}}\right)\cdot V(U(f)) = \left(\prod_{\mathfrak{P}}\prod_{\mathrm{j}=0}^{e_{\mathfrak{P}}-1}\left(\prod_{i=1}^{f_{\mathfrak{P}}}h_{\mathfrak{P},i}^{p+1}-\prod_{i=1}^{f_{\mathfrak{P}}}(\Theta_{\mathfrak{P},i}^{[\mathrm{j}]})^{p-1}\right)\right)(f).$$

The poles of $U(f)$ are supported on the complement of $\mathfrak{M}(k,\mu_N)^{\mathrm{ord}}$ in $\mathfrak{M}(k,\mu_N)$, which is the union of the distinct reduced divisors $W_{\mathfrak{P},i}$ defined by $h_{\mathfrak{P},i}$. It follows from 13.14 that V increases the poles by p. On the other hand the order of vanishing of $\prod_{\mathfrak{P}}\left(\prod_{i=1}^{f_{\mathfrak{P}}}h_{\mathfrak{P},i}^{e_{\mathfrak{P}}(p+1)-a_{\mathfrak{P},i}}\right)$ along any $W_{\mathfrak{P},i}$ is at most $p-1$. Hence, if $U(f)$ is not holomorphic, so is the left hand side. But, the modular form on the right hand side of the equality has no poles by 12.39.

14.8 Remark. The argument above, as well as the case $g=1$ and weight 1 cf. [Gr, §4], suggests that in general one needs to require a condition such as $a_{\mathfrak{P},i}\geq(e_{\mathfrak{P}}-1)p+e_{\mathfrak{P}}+1$ for all $\mathfrak{P}$ and i to guarantee that U takes holomorphic modular forms to holomorphic modular forms.

15 Applications to filtrations of modular forms.

Let $f\in\mathbf{M}(k,\mu_N,\chi)$ be a $\mathfrak{J}$-polarized modular form. Let $\Phi(f)=\prod_{\mathfrak{P},i}\chi_{\mathfrak{P},i}^{a_{\mathfrak{P},i}}$ with $a_{\mathfrak{P},i}\in\mathbf{Z}$ be its filtration as defined in 8.20.

15.1 *A summary.* Let $(\mathfrak{A},\mathfrak{B},\varepsilon,\mathrm{j})$ be an unramified cusp in the sense of 6.4. Suppose that the q-expansion of f at the given cusp is

$$f\Big(\mathbf{Tate}(\mathfrak{A},\mathfrak{B}),\varepsilon,\mathrm{j}\Big)=a_0+\sum_{\nu\in\mathfrak{AB}+}a_\nu q^\nu\in k[\![q^\nu]\!]_{\nu\in\{0\}\cup(\mathfrak{AB})^+}.$$

In the following table we summarize the effect of various operators on the weight χ and the q-expansion of f; the lower part of the table should be understood on the level of q-expansions only.

See 12.38 and 12.40 for the operator $\Theta_{\mathfrak{P},i}^{[\mathrm{j}]}$, 13.1, 13.5 and 13.9 for the operator V, 14.7 for the operator U. Finally, $[h_{\mathfrak{P},i}]$ means multiplication by the partial Hasse invariant $h_{\mathfrak{P},i}$ introduced in 7.12.

Operator	Weight	q-expansion
$[h_{\mathfrak{P},i}](f)$	$\chi\chi_{\mathfrak{P},i-1}^{p}\chi_{\mathfrak{P},i}^{-1}$	$a_0+\sum_\nu a_\nu q^\nu$
$\Theta_{\mathfrak{P},i}^{[\mathrm{j}]}(f)$	$\chi\chi_{\mathfrak{P},i-1}^{p}\chi_{\mathfrak{P},i}$	$\sum_\nu \tilde{\chi}_{\mathfrak{P},i}^{[\mathrm{j}]}(\nu)a_\nu q^\nu$
$V(f)$	$\chi^{(p)}$	$a_0+\sum a_\nu q^{p\nu}$
$U(f)$	χ	$a_0+\sum_\nu a_{p\nu}q^\nu$
$\Lambda(\mathfrak{P},\mathrm{j})(f)$	$\chi \bmod \mathbb{X}_k(1)$	$a_0+\sum_{\nu\in(\mathfrak{P}^{\mathrm{j}+1}\mathfrak{A}\mathfrak{B})^+} a_\nu q^\nu$
$\Lambda(f)$	$\chi \bmod \mathbb{X}_k(1)$	$a_0+\sum a_{p\nu}q^{p\nu}$

15.2 Proposition. *Let $(\mathfrak{A},\mathfrak{B},\varepsilon,\mathrm{j})$ be a $\mathfrak{J}$-polarized unramified cusp of level μ_N. Let $\widetilde{\mathbf{M}}(k,\mu_N)$ be the subring of $k\big((\mathfrak{A},\mathfrak{B},\varepsilon,\mathrm{j})\big)$ generated by the q-expansions at this cusp of modular forms of level μ_N and any weight. Consider the linear operator $\Theta := \mathrm{Id}-\Lambda$ on $\widetilde{\mathbf{M}}(k,\mu_N)$; see 14.1 for the notation. The operator Θ takes χ-eigenfunctions to χ-eigenfunctions and has the following effect on q-expansions: $a_0+\sum_{\nu\in\mathfrak{A}\mathfrak{B}^+} a_\nu q^\nu \mapsto \sum_{\nu\notin p\mathfrak{A}\mathfrak{B}^+} a_\nu q^\nu$.*

The following sequence of k-vector spaces is exact:

$$0 \longrightarrow \widetilde{\mathbf{M}}(k,\mu_N)\xrightarrow{V}\widetilde{\mathbf{M}}(k,\mu_N)\xrightarrow{\Theta}\widetilde{\mathbf{M}}(k,\mu_N)\xrightarrow{U}\widetilde{\mathbf{M}}(k,\mu_N)\longrightarrow 0.$$

Proof: By the table above one sees that $U\circ\Theta=0$ and that $\Theta\circ V=0$. One also concludes also that V is injective and $U\circ V=\mathrm{Id}$. In particular, U is surjective. By definition (see 14.6), we have $V\circ U=\Lambda=\mathrm{Id}-\Theta$. We conclude that if $f\in\mathrm{Ker}(\Theta)$, then $f=(\mathrm{Id}-\Theta)(f)=V\big(U(f)\big)$. Hence, $f\in\mathrm{Im}(V)$. Finally, if $f\in\mathrm{Ker}(U)$, then $f=\sum_{\nu\notin p\mathfrak{A}\mathfrak{B}^+} a_\nu q^\nu$. Hence, $f=\Theta(f)$ i. e., $f\in\mathrm{Im}\big(\Theta\big)$.

15.3 *Caveat.* For $g=1$, the operator Θ defined above is θ^{p-1}, where θ is the classical theta operator as in [Gr, §4].

15.4 Definition. *Let g be a rational function on $\mathfrak{M}$. For every prime $\mathfrak{P}$ over p and any $1\le i\le f_{\mathfrak{P}}$, define*

$$v_{\mathfrak{P},i}(g) := \min\left\{v_C(g)\,|\, C \text{ an irreducible component of } W_{\mathfrak{P},i}\right\}.$$

15.5 Definition. *For any prime $\mathfrak{P}$ above p and any $1\le i\le f_{\mathfrak{P}}$, let*

$$\psi_{\mathfrak{P},i} := \chi_{\mathfrak{P},i-1}^{p}\chi_{\mathfrak{P},i}^{-1}$$

be the weight of the partial Hasse invariant $h_{\mathfrak{P},i}$ defined in 7.12. Define two positive cones in $\mathbb{X}_k\otimes_{\mathbf{Z}}\mathbf{Q}$ by

$$\mathbb{X}_k^+ := \Big\{\prod_{\mathfrak{P},i}\psi_{\mathfrak{P},i}^{a_{\mathfrak{P},i}} : a_{\mathfrak{P},i}\in\mathbf{Q}_{\ge 0}\Big\} \qquad \mathbb{X}_k^{\dagger} = \Big\{\prod_{\mathfrak{P},i}\chi_{\mathfrak{P},i}^{a_{\mathfrak{P},i}} : a_{\mathfrak{P},i}\in\mathbf{Q}_{\ge 0}\Big\},$$

($\mathbb{X}_k^{\dagger}$ is the cone generated by $\mathbb{X}_{O_K}^{\dagger}$, defined in 4.1, under the reduction map). Finally, we define a partial ordering $\le_k$ on $\mathbb{X}_k$ by requiring that if χ, ψ are in $\mathbb{X}_k$,

$$\chi\le_k\psi \iff \psi\chi^{-1}\in\mathbb{X}_k^+.$$

15.6 Lemma. *The elements* $\{\psi_{\mathfrak{P},i}\}_{\mathfrak{P},i}$ *form a basis of* $\mathbb{X}_k \otimes_{\mathbf{Z}} \mathbf{Q}$. *Moreover,*

$$\mathbb{X}_k^{\dagger} \subset \mathbb{X}_k^{+} \qquad \text{and} \qquad \mathbb{X}_k(1) = \{\prod_{\mathfrak{P},i} \psi_{\mathfrak{P},i}^{a_{\mathfrak{P},i}} | a_{\mathfrak{P},i} \in \mathbf{Z}\}.$$

See 4.11 for the definition of $\mathbb{X}_k(1)$.

Proof: For any prime $\mathfrak{P}$ the basic characters $\{\chi_{\mathfrak{P},i}\}_i$ defined in 4.1 are expressed in terms of $\{\psi_{\mathfrak{P},i}\}_i$ by the positive rational $f_{\mathfrak{P}} \times f_{\mathfrak{P}}$-matrix

$$\frac{1}{p^{f_{\mathfrak{P}}} - 1} \begin{pmatrix} 1 & p & p^2 & \dots & p^{f_{\mathfrak{P}}-1} \\ p^{f_{\mathfrak{P}}-1} & 1 & p & \dots & p^{f_{\mathfrak{P}}-2} \\ \vdots & & & & \vdots \\ p & p^2 & p^3 & \dots & 1 \end{pmatrix}.$$

This proves the inclusion. The other statement is Part (3) of 7.14.

15.7 Lemma. *For any positive integer* k *one has,* $\Phi(f^k) = \Phi(f)^k$.

Proof: We assume that f has weight $\Phi(f)$. We certainly have $\Phi(f^k) \leq \Phi(f)^k$. Suppose that the inequality is strict. Then, for some $\mathfrak{P}$ and i, we have that f^k, therefore f, vanishes on every component of $W_{\mathfrak{P},i}$. Hence, $f/h_{\mathfrak{P},i}$ is also holomorphic. This implies that $\Phi(f)$ is strictly less than the weight of f. Contradiction.

15.8 Question. Over the complex numbers, a non-zero Hilbert modular form has weight in $\mathbb{X}_{O_K}^{\dagger}$. In characteristic p this is no longer true as the example of the partial Hasse invariant shows. Note though that the filtration of a partial Hasse invariant is the trivial character $\mathbb{1}$ that lies in $\mathbb{X}_k^{\dagger}$. We therefore ask: *is there an example of a modular form whose filtration is not in* $\mathbb{X}_k^{\dagger}$? *is not in* $\mathbb{X}_k^{+}$?

15.9 Proposition. *Let* f *be a* $\mathfrak{J}$*-polarized modular form of level* μ_N *over* k. *The filtration of* f *and the poles of* $r(f)$ *determine each other by the following relation. If* $\Phi(f) = \prod_{\mathfrak{P},i} \chi_{\mathfrak{P},i}^{a_{\mathfrak{P},i}}$, *then*

$$v_{\mathfrak{P},i}\big(r(f)\big) = -v_{\mathfrak{P},i}\big(a(\chi)\big) = -\sum_{s=0}^{f_{\mathfrak{P}}-1} a_{\mathfrak{P},i+s} p^s. \tag{15.9.1}$$

Proof: Without loss of generality we may assume that the weight of f is equal to its filtration $\Phi(f)$. In particular, f does not vanish identically along any $W_{\mathfrak{P},i}$. Else, by 8.18 the modular form $f/h_{\mathfrak{P},i}$ is holomorphic of weight strictly smaller than $\Phi(f)$. Since for any irreducible component C of $W_{\mathfrak{P},i}$ we have that $v_{\mathfrak{P},i}\big(a(\chi)\big) = v_C\big(a(\chi)\big)$ and for some such component $v_C(f) = 0$, it follows that $v_{\mathfrak{P},i}\big(r(f)\big) = -v_{\mathfrak{P},i}\big(a(\chi)\big)$. By 9.3,

$$\begin{aligned} v_{\mathfrak{P},i}\big(a(\chi)\big) &= \sum_{j=1}^{f_{\mathfrak{P}}} a_{\mathfrak{P},j} v_{\mathfrak{P},i}\big(a(\chi_{\mathfrak{P},j})\big) \\ &= \sum_{s=0}^{f_{\mathfrak{P}}-1} a_{\mathfrak{P},i+s} p^s. \end{aligned}$$

15.10 Proposition. *Let $\mathfrak{P}$ be a prime over p, $1 \leq i \leq f_{\mathfrak{P}}$ and $0 \leq \mathfrak{j} \leq e_{\mathfrak{P}} - 1$. Let $\Theta_{\mathfrak{P},i}^{[\mathfrak{j}]}$ be the operator introduced in 12.38. Suppose that $f \in \mathbf{M}(k, \mu_N, \chi)^{\mathfrak{P},[\mathfrak{j}]}$. Then*

$$\Phi\left(\Theta_{\mathfrak{P},i}^{[\mathfrak{j}]}(f)\right) \leq_k \Phi(f)\chi_{\mathfrak{P},i-1}^p \chi_{\mathfrak{P},i},$$

with equality in the direction $\chi_{\mathfrak{P},i-1}^p \chi_{\mathfrak{P},i}^{-1}$ if and only if $p \nmid a_{\mathfrak{P},i}$.

Proof: We may assume that $\chi = \Phi(f)$. The filtration $\Phi\big(\Theta_{\mathfrak{P},i}^{[\mathfrak{j}]}(f)\big)$ of $\Theta_{\mathfrak{P},i}^{[\mathfrak{j}]}(f)$ is less or equal to its weight $\Phi(f)\chi_{\mathfrak{P},i-1}^p \chi_{\mathfrak{P},i}$. By definition

$$\Theta_{\mathfrak{P},i}^{[\mathfrak{j}]}(f) = \Theta_{\mathfrak{P},i}^{[\mathfrak{j}]}\big(r(f)\big)a(\chi)a\big(\chi_{\mathfrak{P},i}^2\big)h_{\mathfrak{P},i};$$

see 12.38.

The filtration of $\Theta_{\mathfrak{P},i}^{[\mathfrak{j}]}(f)$ in the $\chi_{\mathfrak{P},i-1}^p \chi_{\mathfrak{P},i}^{-1}$-direction is smaller than its weight if and only if for every component C of $W_{\mathfrak{P},i}$ we have

$$v_C\big(\Theta_{\mathfrak{P},i}^{[\mathfrak{j}]}(r(f))\big) \geq v_C(a(\chi)) - v_C(a(\chi_{\mathfrak{P},i}^2)) = v_C(a(\chi)) - 2 \qquad (15.10.1)$$

Since the weight of f is assumed to be equal to its filtration, there is a component C of $W_{\mathfrak{P},i}$ along which f does not vanish. For any such, we have by 15.9

$$v_C(r(f)) = -v_C(a(\chi)) = -\sum_{s=0}^{f_{\mathfrak{P}}-1} a_{\mathfrak{P},i+s}p^s.$$

In particular, $p | v_{\mathfrak{P},i}\big(r(f)\big)$ if and only if $p|a_{\mathfrak{P},i}$. By 12.37, we then get

$$v_C\left(\Theta_{\mathfrak{P},i}^{[\mathfrak{j}]}\big(r(f)\big)\right)\begin{cases} \geq v_C\big(a(\chi)\big) - 2 & \text{if } p|a_{\mathfrak{P},i}; \\ = v_C\big(a(\chi)\big) - 2 - \big(p^{f_{\mathfrak{P}}} - 1\big) & \text{if } p \nmid a_{\mathfrak{P},i}. \end{cases}$$

Thus, (15.10.1) holds if and only if $p|a_{\mathfrak{P},i}$.

To conclude the proof it suffices to prove that (15.10.1) holds for every irreducible component C of $W_{\mathfrak{P},i}$ along which f vanishes. For any such, recalling that we are taking valuations in $\mathfrak{M}^*$, one has

$$v_C(r(f)) \geq -v_C(a(\chi)) + (p^{f_{\mathfrak{P}}} - 1).$$

As above we compute that

$$v_C\left(\Theta_{\mathfrak{P},i}^{[\mathfrak{j}]}\big(r(f)\big)\right) \geq v_C(r(f)) - 2 - (p^{f_{\mathfrak{P}}} - 1) \geq -v_C(a(\chi)) - 2$$

as claimed.

15.11 Remark. Since $\big(\Theta_{\mathfrak{P},i}^{[\mathfrak{j}]}\big)^p = \Theta_{\mathfrak{P},i+1}^{[\mathfrak{j}]}$ one can not hope to strengthen 15.10 i. e., that equality holds if $p \nmid a_{\mathfrak{P},i}$.

15.12 Proposition. *See 13.1 and 13.5 for the definition of the operator V and its effect on weights. We have $\Phi\big(V(f)\big) = \Phi(f)^{(p)}$.*

Proof: By definition V is induced by Frobenius on $\mathfrak{M}(k, \mu_N)$. In particular, it increases the poles or the zeroes of f by p. Hence, if some partial Hasse invariant divides $V(f)$, it must divide f itself.

15.13 Proposition. *Let f be a $\mathfrak{J}$-polarized Hilbert modular form of level μ_N and filtration $\Phi(f) = \prod_{\mathfrak{P},i} \chi_{\mathfrak{P},i}^{a_{\mathfrak{P},i}}$ such that $U(f)$ is also holomorphic e. g., $a_{\mathfrak{P},i} \geq 2$ for every $\mathfrak{P}$ and every i. Then,*

$$\Phi\big(U(f)\big)^{(p)} \leq_k \Phi(f)\mathbf{Nm}^{p^2-1},$$

with strict inequality if p is ramified or $a_{\mathfrak{P},i} \not\equiv 1$ modulo p for some $\mathfrak{P}$ and i.

Proof: By (14.7.1), we have

$$h^{p+1}V\big(U(f)\big) = \prod_{\mathfrak{P}} \prod_{i=1}^{f_{\mathfrak{P}}} h_{\mathfrak{P},i}^{a_{\mathfrak{P},i}} \cdot \prod_{\mathfrak{P}} \prod_{\mathrm{j}=0}^{e_{\mathfrak{P}}-1} \Big(\prod_{i=1}^{f_{\mathfrak{P}}} h_{\mathfrak{P},i}^{p+1} - \prod_{i=1}^{f_{\mathfrak{P}}} (\Theta_{\mathfrak{P},i}^{[\mathrm{j}]})^{p-1}\Big)(f).$$

By 15.12 $V\big(U(f)\big)$ has filtration $\Phi\big(U(f)\big)^{(p)}$. For each $\mathbf{j} \in \mathbf{N}$, each prime $\mathfrak{P}$ and each $g \in \mathbf{M}\big(k, \mu_N, \chi\big)$ we have

$$\Phi\left(\Big(\prod_{i=1}^{f_{\mathfrak{P}}} h_{\mathfrak{P},i}^{p+1} - \prod_{i=1}^{f_{\mathfrak{P}}} (\Theta_{\mathfrak{P},i}^{[\mathrm{j}]})^{p-1}\Big)(g)\right) \leq \sup\left\{\Phi(g), \Phi\Big(\prod_{i=1}^{f_{\mathfrak{P}}} ((\Theta_{\mathfrak{P},i}^{[\mathrm{j}]})^{p-1})(g)\Big)\right\}.$$

By 15.10 we have

$$\Phi\Big(\prod_{i=1}^{f_{\mathfrak{P}}} (\Theta_{\mathfrak{P},i}^{[\mathrm{j}]})^{p-1}(g)\Big) \leq_k \Phi(g) \prod_{i=1}^{f_{\mathfrak{P}}} \chi_{\mathfrak{P},i}^{p^2-1}.$$

This proves the first part the Proposition. By 15.10 we have strict inequality if there exist $\mathfrak{P}$, i, $0 \leq j < e_{\mathfrak{P}}(p-1)$ such that $a_{\mathfrak{P},i} + j \equiv 0 \bmod p$ i. e., either p is ramified or $a_{\mathfrak{P},i} \not\equiv 1 \bmod p$.

15.14 Definition. *We say that f is an ordinary form if there exists $\lambda \in k^*$ such that $U(f) = \lambda f$.*

Note that in the definition of an ordinary form f, we do not require f to be an eigenform for the Hecke operators.

15.15 Corollary. *The notation is as in 15.13. If $a_{\mathfrak{P},i} > e_{\mathfrak{P}}(p+1)$ for some $\mathfrak{P}$ and i, then $\Phi\big(U(f)\big) <_k \Phi(f)$.*

In particular, if f is an ordinary form and $\chi = \prod_{\mathfrak{P},i} \chi_{\mathfrak{P},i}^{a_{\mathfrak{P},i}}$ is its filtration, then $a_{\mathfrak{P},i} \leq e_{\mathfrak{P}}(p+1)$ for every $\mathfrak{P}$ and i.

Proof: By hypothesis, $\Phi(f)^{(p)-1} = \prod_{\mathfrak{P},i} \psi_{\mathfrak{P},i}^{a_{\mathfrak{P},i}} >_k \prod_{\mathfrak{P},i} \psi_{\mathfrak{P},i}^{e_{\mathfrak{P}}(p+1)} = \mathbf{Nm}^{p^2-1}$; see 15.5 for the notation $\psi_{\mathfrak{P},i}$. Therefore, $\Phi(f)^{(p)} >_k \Phi(f)\,\mathbf{Nm}^{p^2-1}$. Using 15.13 we get that $\Phi(f)^{(p)} >_k \Phi\big(U(f)\big)^{(p)}$. Since the operation $\chi \mapsto \chi^{(p)}$ induces a bijection on the positive cone $\mathbb{X}_k^+$ we conclude that $\Phi\big(U(f)\big) <_k \Phi(f)$.

15.16 Remark. The assumption $a_{\mathfrak{P},i} > e_{\mathfrak{P}}(p+1)$ for all $\mathfrak{P}$ and i, implies that $\Phi\big(U(f)\big) <_k \Phi(f)$ with respect to any $\psi_{\mathfrak{P},i}$.

15.17 Remark. For every prime $\mathfrak{P}$ and $1 \leq i \leq f_{\mathfrak{P}}$ we can ensure, by multiplying by suitable positive powers of partial Hasse invariants, that $2 \leq a_{\mathfrak{P},j} \leq e_{\mathfrak{P}}(p+1)$ except possibly for $j = i$. In order to get a result of the type $a_{\mathfrak{P},j} \in [t, \dots, e_{\mathfrak{P}}(p+1)]$

for *every* $\mathfrak{P}, j$ (for some t, preferably $t = 0$ or 1), as in the $g = 1$ case, one needs a positivity result for filtrations of Hilbert modular forms mod p; cf. 15.8.

16 Theta cycles and parallel filtration (inert case).

In this section we deal with two further phenomena concerning modular forms, under the assumption that p is inert in L. The first is the case of modular forms of parallel weight. Those are of particular interest from the point of view of Galois representations and it makes sense to discuss a "parallel theory" for them. In fact, due to the ampleness of the Hodge bundle, we are able to improve on our results for modular forms of non-parallel weight; cf. 15.17.

The second topic we take is the theory of theta cycles, for not necessarily parallel weight. So far, we are only able to present some preliminary results that indicate that the behavior of theta cycles is much more complicated than in the elliptic case. The subject definitely deserves further study and the authors hope to discuss it in greater depth on a future occasion.

Throughout this section we assume that p is inert in O_L i. e., that $\mathfrak{P} := pO_L$ is a prime ideal of O_L. For $i = 1, \ldots, g$ we write χ_i for the basic weight $\chi_{\mathfrak{P},i}$, h_i for the partial Hasse invariant $h_{\mathfrak{P},i}$ defined in 7.12 and $\psi_i = \chi_{i-1}^p \chi_i^{-1}$ for its weight.

16.1 Definition. *A $\mathfrak{J}$-polarized Hilbert modular form over k and of level μ_N is said to be of parallel weight if its weight is a power of* $\mathbf{Nm}$.

As an example, note that the total Hasse invariant $h := \prod_i h_i$ is a modular form of parallel weight $\mathbf{Nm}^{p-1}$.

16.2 *The operators θ, V and U.* If f is a modular form of parallel weight, define $V(f)$ as in 13.1, $U(f)$ as in 14.7 and

$$\theta(f) := (\Theta_1 \circ \cdots \circ \Theta_g)(f).$$

By 12.39, 13.5 and 14.7, the operators θ, V and U preserve the space of modular forms of parallel weight. Furthermore, θ and V send holomorphic modular forms to holomorphic modular forms and U sends holomorphic modular forms of weight $\mathbf{Nm}^c$ to holomorphic modular forms of the same weight if $c \geq 2$.

Let $f \in \mathbf{M}(k, \mu_N, \mathbf{Nm}^c)$. Let $(\mathfrak{A}, \mathfrak{B}, \varepsilon, \mathrm{j})$ be an unramified cusp in the sense of 6.4 and let

$$f\Big(\mathbf{Tate}(\mathfrak{A}, \mathfrak{B}), \varepsilon, \mathrm{j}\Big) = a_0 + \sum_{\nu \in \mathfrak{A}\mathfrak{B}^+} a_\nu q^\nu \in k[\![q^\nu]\!]_{\nu \in \{0\} \cup (\mathfrak{A}\mathfrak{B})^+}.$$

The following table summarizes the effects of the above operators on q-expansions and weights.

Operator	Power of $\mathbf{Nm}$	q-expansion
$[h](f)$	$c+p-1$	$a_0 + \sum_\nu a_\nu q^\nu$
$\theta(f)$	$c+p+1$	$\sum_\nu \widetilde{\mathbf{Nm}}(\nu) a_\nu q^\nu$
$V(f)$	$p \cdot c$	$a_0 + \sum a_\nu q^{p\nu}$
$U(f)$	c	$a_0 + \sum_\nu a_{p\nu} q^\nu$

Here, $\widetilde{\mathbf{Nm}} := \prod_i \tilde{\chi}_i(\nu)$ and $[h](f) := h \cdot f$. Furthermore, we have the classical identity between modular forms (deduced from (14.7.1))

$$h^{p+1} \cdot VU(f) = h^c \cdot \left((h^{p+1} - \theta^{p-1})(f)\right) \tag{16.2.1}$$

($VU = \mathrm{Id} - \theta^{p-1}$ on q-expansions).

16.3 Remark. Given two modular forms f and f' of parallel weights χ and χ', we have $\chi \leq_k \chi'$ in the sense of 15.5 if and only if $\chi'\chi^{-1} = \mathbf{Nm}^a$ with $a \in \mathbf{N}$.

16.4 Lemma. *Let $f \in \mathbf{M}(k, \mu_N, \mathbf{Nm}^c)$. If $f \neq 0$, then $c \geq 0$.*

Proof: Denote by $\pi \colon A \to \mathfrak{M}(k, \mu_N)$ the universal abelian scheme over $\mathfrak{M}(k, \mu_N)$. Then, $\mathbf{M}(k, \mu_N, \mathbf{Nm}^c)$ is the space of global sections of the c power of the determinant of the Hodge bundle $\pi_* \Omega^1_{A/\mathfrak{M}(k,\mu_N}$. Since det $\pi_* \Omega^1_{A/\mathfrak{M}(k,\mu_N}$ is ample, the conclusion follows.

16.5 Definition. *Let $f \in \mathbf{M}(k, \mu_N, \mathbf{Nm}^c)$ be a $\mathfrak{I}$-polarized Hilbert modular form over k, of level μ_N and of parallel weight c. Define the parallel filtration of f, denoted by $\Phi^{||}(f)$, to be the minimal parallel weight for which there exists a $\mathfrak{I}$-polarized modular form over k, of level μ_N, having the same q-expansion as f at some (hence, any) cusp.*

16.6 Proposition. *The notation is as in 16.5. There is a unique $\mathfrak{I}$-polarized modular form $t^{||}$ over k of level μ_N, of weight $\Phi^{||}(f)$ and with the same q-expansion as f. If t is a $\mathfrak{I}$-polarized modular form over k of level μ_N, of parallel weight and with the same q-expansion of f, then there exist non-negative integer b such that $t = t^{||} h^b$.*

Proof: The existence of $t^{||}$ follows from the definition of $\Phi^{||}(f)$. By the q-expansion principle, two modular forms of the same weight and q-expansion at a cusp are equal. This proves the uniqueness of $t^{||}$. By 7.14 and 7.22, two modular forms of parallel weight having the same q-expansion at some cusp differ by a power of the total Hasse invariant.

16.7 Proposition. *Let $f \in \mathbf{M}(k, \mu_N, \mathbf{Nm}^c)$. Let $\Phi(f) = \chi_1^{a_1} \dots \chi_g^{a_g}$ be the filtration of f as in 8.20. Let*

$$r_i := \frac{p-1}{p^g - 1} \sum_{s=0}^{g-1} a_{i+s} p^s, \qquad r := \max\{r_1, \dots, r_g\}.$$

Then, we have

$$\Phi(f) = \psi_1^{\frac{r_1}{p-1}} \cdots \psi_g^{\frac{r_g}{p-1}}, \qquad \Phi^{||}(f) = \mathbf{Nm}^r.$$

Let $t^{||}$ (respectively, t) be a modular form of weight $\Phi^{||}(f)$ (resp. $\Phi(f)$) and with the same q-expansion as f, then

$$t^{||} = t \cdot \prod_i h_i^{\frac{r - r_i}{p-1}}.$$

In particular, $r_{i_0} = r$ if and only if $t^{||}$ does not vanish on the zero locus W_{i_0} of h_{i_0}.

Proof: The formula for $\Phi(f)$ follows by applying the matrix transforming the basis of $\mathbb{X}_k \otimes \mathbf{Q}$ given by the basic characters $\{\chi_1, \ldots, \chi_g\}$ into the basis $\{\psi_1, \ldots, \psi_g\}$; see 15.6. Write $\Phi^{\|}(f) = \mathbf{Nm}^w$. By definition of filtration we may write $t^{\|} = t \cdot \prod_i h_i^{b_i}$ with $b_i \in \mathbf{N}$. Comparing weights, we have $\prod_i \psi_i^{\frac{w}{p-1}} = \prod_i \psi_i^{\frac{r_i}{p-1}+b_i}$. Since $t^{\|}$ does not vanish identically on the zero locus of h, there exists $1 \leq i_0 \leq g$ such that $b_{i_0} = 0$. Then,

$$w = r_{i_0} = \frac{p-1}{p^g-1} \sum_{s=0}^{g-1} a_{i_0+s} p^s.$$

For every i, $w = r_i + (p-1)b_i \geq r_i$. This implies that $w = r = r_{i_0}$. Furthermore, $(p-1)b_i = r - r_i$, which gives $t^{\|} = t \cdot \prod_i h_i^{\frac{r-r_i}{p-1}}$. Thus, $t^{\|}$ vanishes along W_i if and only if $r > r_i$.

16.8 Corollary. *For any $a \in \mathbf{N}$ we have $\Phi^{\|}(f^a) = \Phi^{\|}(f)^a$.*

16.9 Corollary. *We have $\Phi^{\|}\big(V(f)\big) = \Phi^{\|}(f)^p$.*

16.10 Proposition. *Let $f \in \mathbf{M}\big(k, \mu_N, \mathbf{Nm}^c\big)$. We have,*

$$\Phi^{\|}(\theta(f)) \leq_k \Phi^{\|}(f)\mathbf{Nm}^{p+1}.$$

Furthermore, if $p|\Phi^{\|}(f)$, then $\Phi^{\|}(\theta(f)) <_k \Phi^{\|}(f)\mathbf{Nm}^{p+1}$.

Proof: The first statement is clear. In the notation of 16.7, write $\Phi(f) = \prod_i \psi_i^{\frac{r_i}{p-1}}$. Let $\{i_0, \ldots, i_s\}$ be the set of indices for which $r_{i_j} = r$. Then, $p|r$ implies that $p|a_{i_j}$. By 15.10, we have that $\Phi(\theta(f)) \leq_k \prod_i \psi_i^{\frac{r_i+p+1}{p-1}} \prod_j \psi_{i_j}^{-1}$. Write $\Phi(\theta(f)) = \prod_i \psi_i^{\frac{r_i'}{p-1}}$. Then, $r_i' \leq r_i + p + 1$ for every $1 \leq i \leq g$ and $r_{i_j}' < r_{i_j} + p + 1$ for $j = 0, \ldots, s$. Hence, $\max\{r_i' | i = 1, \ldots, g\} < r + p + 1$. The second statement now follows from 16.7.

16.11 Remark. Contrary to the elliptic case, it seems that in general the assumption $p \not| \Phi^{\|}(f)$ does not imply that $\Phi^{\|}(\theta(f)) = \Phi^{\|}(f)\mathbf{Nm}^{p+1}$.

16.12 Proposition. *Let f be a $\mathfrak{I}$-polarized Hilbert modular form of level μ_N, parallel weight and parallel filtration $\Phi^{\|}(f) = \mathbf{Nm}^r$. Assume that $U(f)$ is holomorphic e. g., $r \geq 2$. Then,*

$$\Phi^{\|}\big(U(f)\big)^p \leq_k \Phi^{\|}(f)\mathbf{Nm}^{p^2-1},$$

with strict inequality if $p|r$.

Proof: We get from (16.2.1) that $h^{p+1}VU(f) = h^c\big((h^{p+1} - \theta^{p-1})(f)\big)$. Since $\Phi^{\|}(VU(f)) = \Phi^{\|}(U(f))^p$ by 16.9, we conclude using 16.10.

16.13 Proposition. *Let f be a form of parallel weight. If $\Phi^{\|}(f) > \mathbf{Nm}^{p+1}$, then $\Phi^{\|}\big(U(f)\big) <_k \Phi^{\|}(f)$.*

Moreover, if f is an ordinary form of parallel weight, there exists an ordinary form f' of weight $\mathbf{Nm}^{r'}$, with the same q-expansion as f, such that $2 \leq r' \leq p+1$.

Proof: Let $\mathbf{Nm}^r = \Phi^{\|}(f)$. Since $r > p+1$, we have by 16.12

$$\Phi^{\|}(f)^p = \mathbf{Nm}^{pr} >_k \mathbf{Nm}^r \mathbf{Nm}^{p^2-1} = \Phi^{\|}(f)\mathbf{Nm}^{p^2-1} \geq_k \Phi^{\|}(U(f))^p,$$

and the first assertion follows.

To prove the second claim, note that $0 \leq \Phi^{\|}(f) \leq p+1$ by the first statement and 16.4. If $\Phi^{\|}(f) = 0, 1$ and $t^{\|}$ is as in 16.6, then $f' := t^{\|}h$ has the required properties.

We next discuss, following [Jo], theta cycles and until the end of this section we work with modular forms whose weight is not necessarily parallel. From the point of view of understanding the filtrations of q-expansions obtained by "twists", i.e., by applications of the theta operators $\Theta_1, \dots, \Theta_g$, it is enough to consider only powers of a single theta operator, say Θ_1.

16.14 *Θ_1-cycles.* Let f be a $\mathfrak{J}$-polarized Hilbert modular form of level μ_N and weight χ over k. Then, $U(f) = 0$ if and only if $\Theta_1^{p^g-1}(f)$ and f have the same q-expansion. In that case, we say that f *belongs to its Θ_1-cycle.* In particular, $\Phi(\Theta_1^{p^g-1}(f)) = \Phi(f)$.

16.15 Definition. *Let f be a $\mathfrak{J}$-polarized Hilbert modular in* $\mathrm{Ker}(U)$ *of weight χ. We call the set $\{\Phi(\Theta_1^A(f)) | A = 0, \dots p^g - 2\}$ the Θ_1-cycle belonging to f.*

For $0 \leq A \leq p^g - 2$, we say that A is a low point of the Θ_1-cycle of f if $\Phi(\Theta^A(f)) <_k \Phi(\Theta^{A-1}(f))\chi_g^p\chi_1$ in the $\psi_1 = \chi_g^p\chi_1^{-1}$ direction i. e., $\Phi(\Theta^A(f)) \leq_k \Phi(\Theta^{A-1}(f))\chi_1^2$.

Assume that $\Phi(f) = \chi$ and, without loss of generality, that 0 is a low point of the Θ_1-cycle of f. Label by $0 = A_0 < A_1 < \ldots < A_s$ the low points of the Θ_1-cycle of f. For every $i = 0, \dots, s$ put

$$c_i := A_{i+1} - A_i,$$

where by c_s we mean $p^g - 1 - A_s$. Let b_i be the positive integer defined by

$$\Phi(\Theta^{A_i-1}(f)) \cdot \chi_g^p\chi_1 = \Phi(\Theta^{A_i}(f)) \cdot \psi_1^{b_i} \prod_{j\neq 1} \psi_j^{\alpha_{i,j}}$$

with $\alpha_{i,j} \in \mathbf{N}$.

16.16 Proposition. *The following hold:*

a) *$\sum_{i=0}^s c_i = p^g - 1$ and $0 < c_i < p$ for every i;*

b) *$\sum_{i=0}^s b_i = p^g + 1$;*

c) *$c_i + b_i \equiv 0 \bmod p$.*

Proof: The length of the Θ_1-cycle is $p^g - 1$. Hence, $\sum_{i=0}^s c_i = p^g - 1$. By 15.10, for every modular form g of filtration $\prod_i \chi_i^{a_i}$ the filtration of $\Theta_1(g)$ is strictly less in the ψ_1 direction than $\Phi(g)\chi_g^p\chi_1$ if and only if $p|a_1$. Thus, if $g = \Theta_1^{A_i}(f)$ after iterating Θ_1 at most p times another low point must appear. Claim a) follows.

Using the matrix in 15.6, one finds that $\chi_g^p\chi_1 = \psi_1^{\frac{p^g+1}{p^g-1}} \cdot \psi_2^* \cdots \psi_g^*$. Therefore, up to multiplication by (rational) powers of $\psi_2, \dots, \psi_g$, we have $\Phi(\Theta^{A_i}(f)) =$

$\Phi(\Theta^{A_i-1}(f))\psi_1^{\frac{p^g+1}{p^g-1}-b_i}$ and $\Phi(\Theta^j(f)) = \Phi(\Theta^{j-1}(f))\psi_1^{\frac{p^g+1}{p^g-1}}$ for $j \neq A_i$. Since f belongs to its Θ_1-cycle, $\Phi(\Theta_1^{p^g-1}(f)) = \Phi(f)$ and we conclude that $(p^g-1)\frac{p^g+1}{p^g-1} = \sum_i b_i$. Claim b) follows.

Let $\Phi = \Phi(\Theta^{A_i-1}(f)) = \prod_j \chi_j^{a_j}$. Using 15.10 and the assumption that A_i is a low point, we have $p|a_1$. In the notation above, $\Phi(\Theta_1^{A_i}(f)) = \Phi\chi_1^{b_i+1}$ modulo the subgroup $H := \langle\chi_1^p, \chi_2, \ldots, \chi_g\rangle$. Modulo H the filtration of $\Theta_1^{A_{i+1}-1}(f) = \Theta_1^{c_i-1}(\Theta_1^{A_i}(f))$ is $\Phi\chi_1^{b_i+1}\cdot\chi_1^{c_i-1}$. Since A_{i+1} is a low point, we have that $a_1+(b_i+1)+(c_i-1) \equiv 0 \bmod p$. Claim c) follows.

16.17 Remark. The Proposition shows that there are at least $\frac{p^g-1}{p-1}$ low points in the Θ_1-cycle of f. Also, note that the weight $\mathrm{wt}(\Theta_1^{p^g-1}(f))$ of $\Theta_1^{p^g-1}(f)$ is equal to $\mathrm{wt}(f)(\chi_g^p\chi_1)^{p^g-1}$, while $\Phi(\Theta_1^{p^g-1}(f)) = \Phi(f) = \mathrm{wt}(f)$. Using the identity $(\chi_g^p\chi_1)^{p^g-1} = \psi_1^{p^g+1}\psi_2^{2p^{g-1}}\cdots\psi_g^{2p}$, we conclude that there must be drops in the $\psi_2,\ldots,\psi_g$ directions along the Θ_1-cycle of f (the accumulated drop in the ψ_i direction being $\psi_i^{2p^{g-i+1}}$ for $1 < i \leq g$). The position of these drops along the Θ_1-cycle seems mysterious at present. Thus, contrary to the elliptic case studied in [Jo], the combinatorics of the Θ_1-cycle seems to be quite complicated.

17 Functorialities.

Let $L_1 \subset L_2$ be an extension of totally real number fields. Let $\mathfrak{I}_i \subset L_i$ be fixed fractional ideals such that $\mathfrak{I}_2 = \mathfrak{I}_1 \otimes_{O_{L_1}} \mathrm{D}_{L_2/L_1}$. We add subscripts 1 or 2 to the usual notations for the degrees over $\mathbf{Q}$, the primes, the ramification indices, the degrees of the residue fields, the associated moduli spaces and spaces of modular forms.

17.1 Definition. *Let N be an integer and let S be a scheme. We define a morphism*

$$\Upsilon\colon \mathfrak{M}_1(S,\mu_N) \longrightarrow \mathfrak{M}_2(S,\mu_N),$$

where $\mathfrak{M}_i(S,\mu_N)$ is the moduli space with respect to $\mathfrak{I}_i$-polarization. Consider an abelian scheme $(A_1,\iota_1,\lambda_1,\varepsilon_1)$ over a S-scheme T, with real multiplication ι_1 by O_{L_1}, polarization type $\lambda_1\colon (M_{A_1}, M_{A_1}^+)\xrightarrow{\sim}(\mathfrak{I}_1,\mathfrak{I}_1^+)$ and μ_N-level structure ε_1; see 3.2. Then

$$\Upsilon(A_1,\iota_1,\lambda_1,\varepsilon_1) := (A_2,\iota_2,\lambda_2,\varepsilon_2),$$

where

$$A_2 := A_1 \underset{O_{L_1}}{\otimes} \mathrm{D}_{L_2/L_1}^{-1} \to T$$

is an abelian scheme over T with real multiplication

$$\iota_2 := \iota_1 \otimes \mathrm{id}\colon O_{L_2} \hookrightarrow \mathrm{End}_T(A \underset{O_{L_1}}{\otimes} \mathrm{D}_{L_2/L_1}^{-1}) = \mathrm{End}_T(A_2)$$

and μ_N-level structure

$$\varepsilon_2 := \varepsilon_1 \otimes \mathrm{id}\colon \mu_N \underset{\mathbf{Z}}{\otimes} \mathrm{D}_{L_1}^{-1} \underset{O_{L_1}}{\otimes} \mathrm{D}_{L_2/L_1}^{-1} \hookrightarrow A_1 \underset{O_{L_1}}{\otimes} \mathrm{D}_{L_2/L_1}^{-1} = A_2.$$

Note that

$$(A_2)^\vee := \mathrm{Ext}_T^1(A_2,\mathbf{G}_{m,T}) \xleftarrow{\sim} \mathrm{Ext}_T^1(A_1,\mathbf{G}_{m,T}) \underset{O_{L_1}}{\otimes} O_{L_2} \xrightarrow{\sim} A_1^\vee \underset{O_{L_1}}{\otimes} O_{L_2},$$

where $\vee$ denotes the dual abelian scheme. Hence, we get the polarization type

$$\lambda_2 := \lambda_1 \otimes \mathrm{id}\colon \left(M_{A_2}, M_{A_2}^+\right) \xrightarrow{\sim} \left(\mathfrak{I}_2, \mathfrak{I}_2^+\right)$$

with $\mathfrak{I}_2 := \mathfrak{I}_1 \otimes_{O_{L_1}} \mathrm{D}_{L_2/L_1}$.

17.2 *The extension of Υ to the cusps.* Let $i = 1$ or 2. Let $(\mathfrak{A}_i, \mathfrak{B}_i)$ be two fractional ideals of L_i such that $\mathfrak{A}_i \mathfrak{B}_i^{-1} = \mathfrak{I}_i$. Fix a rational polyhedral cone decomposition $\{\sigma_{i,\beta}\}_\beta$ of the dual cone to $(\mathfrak{A}_i \mathfrak{B}_i)_{\mathbf{R}}^+ \subset (\mathfrak{A}_i \mathfrak{B}_i)_{\mathbf{R}}$ which is invariant under the action of the totally positive units of O_{L_i} and such that, modulo this action, the number of polyhedra is finite. Let

$$S_i := (\mathfrak{A}_i \mathfrak{B}_i)^\vee \underset{\mathbf{Z}}{\otimes} \mathbf{G}_{m,\mathbf{Z}}.$$

We have constructed in 6.3 Tate objects

$$\mathbf{Tate}(\mathfrak{A}_i, \mathfrak{B}_i)_{\sigma_{i,\beta}} = \left(\mathfrak{A}_i^{-1} \mathrm{D}_{L_i}^{-1} \underset{\mathbf{Z}}{\otimes} \mathbf{G}_{m, S_{i,\sigma_{i,\beta}}^\wedge} / \underline{q}(\mathfrak{B}_i)\right) \underset{S_{i,\sigma_{i,\beta}}^\wedge}{\times} \left(S_{i,\sigma_{i,\beta}}^\wedge \backslash S_{i,\sigma_{i,\beta},0}\right).$$

This is an abelian scheme with real multiplication by O_{L_i} over the open subscheme $S_{i,\sigma_{i,\beta}}^\wedge \backslash S_{i,\sigma_{i,\beta},0}$ of $S_{i,\sigma_{i,\beta}}^\wedge$. The latter is defined as the spectrum of the ring obtained completing the affine scheme $S_{i,\sigma_{i,\beta}}$ along the closed subscheme $S_{i,\sigma_{i,\beta},0} = S_{\sigma_{i,\beta}} \backslash S_i$ with reduced structure.

By 6.2 we have that

$$\Upsilon\left(\mathbf{Tate}(\mathfrak{A}_1, \mathfrak{B}_1)_{\sigma_{1,\beta}}\right) \xrightarrow{\sim}$$

$$\left(\mathfrak{A}_1^{-1} \mathrm{D}_{L_2}^{-1} \underset{\mathbf{Z}}{\otimes} \mathbf{G}_{m, S_{1,\sigma_{1,\beta}}^\wedge} / \underline{q}\left(\mathfrak{B}_1 \mathrm{D}_{L_2/L_1}^{-1}\right)\right) \underset{S_{1,\sigma_{1,\beta}}^\wedge}{\times} \left(S_{1,\sigma_{1,\beta}}^\wedge \backslash S_{1,\sigma_{1,\beta},0}\right).$$

Assume that $\mathfrak{A}_2 = \mathfrak{A}_1 O_{L_2}$ and $\mathfrak{B}_2 = \mathfrak{B}_1 \mathrm{D}_{L_2/L_1}^{-1}$. The trace map defines an O_{L_1}-linear, surjective homomorphism $1 \otimes \mathrm{Tr}_{L_2/L_1}\colon \mathfrak{A}_2 \mathfrak{B}_2 \to \mathfrak{A}_1 \mathfrak{B}_1$ and, hence, a closed immersion $S_1 \hookrightarrow S_2$. Choose cone decompositions $\{\sigma_{i,\beta}\}_\beta$ for $i = 1, 2$ as above such that $\{\sigma_{1,\beta}\}_\beta$ is induced by $\{\sigma_{2,\beta}\}_\beta$ via $1 \otimes \mathrm{Tr}_{L_2/L_1}$ for each index β. We get an induced a closed immersion $\rho_\beta\colon S_{1,\sigma_{1,\beta}}^\wedge \hookrightarrow S_{2,\sigma_{2,\gamma}}^\wedge$ such that $\rho_\beta^{-1}\left(S_{2,\sigma_{2,\beta},0}\right) = S_{1,\sigma_{1,\beta},0}$ and

$$\Upsilon\left(\mathbf{Tate}(\mathfrak{A}_1, \mathfrak{B}_1)_{\sigma_{1,\beta}}\right) = \mathbf{Tate}(\mathfrak{A}_2, \mathfrak{B}_2)_{\sigma_{2,\beta}} \underset{S_{2,\sigma_{2,\beta}}}{\times} S_{1,\sigma_{1,\beta}}.$$

17.3 *The canonical homomorphism on weights.* There is a canonical injective homomorphism of $\mathbf{Z}$-group schemes $\mathcal{G}_1 \longrightarrow \mathcal{G}_2$ defined on R-valued points by the inclusion $\left(O_{L_1} \otimes_{\mathbf{Z}} R\right)^* \hookrightarrow \left(O_{L_2} \otimes_{\mathbf{Z}} R\right)^*$. This induces for any scheme T a map of characters defined over T:

$$\Psi\colon \mathbb{X}_T(\mathcal{G}_2) \longrightarrow \mathbb{X}_T(\mathcal{G}_1).$$

It follows from 7.4 that it sends basic (resp. universal) characters to basic (resp. universal) characters.

17.4 *The effect of Υ on modular forms.* Let $f \in \mathbf{M}_2(S, \mu_N, \chi)$ be a modular form over S of level μ_N and weight χ. Define

$$\Upsilon^*(f) \in \mathbf{M}_1\big(S, \mu_N, \Psi(\chi)\big)$$

by requiring that for any affine S-scheme $\mathrm{Spec}(R)$, any abelian scheme A_1 with RM by O_{L_1} and level μ_N and any generator ω of $\mathrm{H}^0(A_1, \Omega^1_{A_1/R})$ as $R \otimes_{\mathbf{Z}} O_{L_1}$-module

$$\Upsilon^*(f)\big(A_1, \iota_1, \lambda_1, \varepsilon_1, \omega\big) := f\left(A_1 \underset{O_{L_1}}{\otimes} \mathrm{D}^{-1}_{L_2/L_1}, \iota_2, \lambda_2, \varepsilon_2, \omega \otimes 1\right).$$

See 17.1 for the notation. Note that

$$\Omega^1_{A_1 \otimes_{O_{L_1}} \mathrm{D}^{-1}_{L_2/L_1}/R} \xrightarrow{\sim} \Omega^1_{A_1/R} \underset{O_{L_1}}{\otimes} O_{L_2}$$

and, hence, $\omega \otimes 1$ is a generator of $\Omega^1_{A_1 \otimes_{O_{L_1}} \mathrm{D}^{-1}_{L_2/L_1}/R}$. One checks that the definition is well posed.

17.5 Lemma. *(The effect of Υ on q-expansions) Let $f \in \mathbf{M}_2(S, \mu_N, \chi)$ and let $(\mathfrak{A}, \mathfrak{B}, \varepsilon, \mathrm{j})$ be a $\mathfrak{I}_1$-polarized unramified cusp of $\mathfrak{M}_1(S, \mu_N)$. Suppose that the q-expansion of f at the cusp $\big(\mathfrak{A}\, O_{L_2}, \mathfrak{B}\, \mathrm{D}^{-1}_{L_2/L_1}, \varepsilon, \mathrm{j}\big)$, in the sense of 6.6, is*

$$f\big(\mathbf{Tate}(\mathfrak{A}\, O_{L_2}, \mathfrak{B}\, \mathrm{D}^{-1}_{L_2/L_1}), \varepsilon, \mathrm{j}\big) = a_0 + \sum_{\nu \in (\mathfrak{A}\mathfrak{B}\mathrm{D}^{-1}_{L_2/L_1})^+} a_\nu q^\nu.$$

Then the q-expansion of $\Upsilon^*(f)$ at the cusp $(\mathfrak{A}, \mathfrak{B}, \mathrm{j}, \varepsilon)$ is

$$a_0 + \sum_{\delta \in (\mathfrak{A}\mathfrak{B})^+} \Big(\sum_{\nu | \mathrm{Tr}_{L_2/L_1}(\nu) = \delta} a_\nu \Big) q^\delta.$$

Proof: We calculate from the definitions

$$\begin{aligned}
\Upsilon^*(f)\big(\mathbf{Tate}(\mathfrak{A}, \mathfrak{B}), \varepsilon, \mathrm{j}\big) &:= \Upsilon^*(f)\Big(\mathbf{Tate}(\mathfrak{A}, \mathfrak{B})_{\sigma_{1,\beta}}, \varepsilon, \frac{dt}{t}\Big) \\
&= f\Big(\mathbf{Tate}(\mathfrak{A}, \mathfrak{B})_{\sigma_{1,\beta}} \underset{O_{L_1}}{\otimes} \mathrm{D}^{-1}_{L_2/L_1}, \varepsilon, \frac{dt}{t} \otimes 1\Big) \\
&= f\Big(\mathbf{Tate}(\mathfrak{A}O_{L_2}, \mathfrak{B}\mathrm{D}^{-1}_{L_2/L_1})_{\sigma_{2,\beta}} \underset{S_{2,\sigma_{2,\beta}}, \rho_\beta}{\times} S_{1,\sigma_{1,\beta}}, \varepsilon, \frac{dt}{t}\Big) \\
&= \rho_\beta\Big(f\Big(\mathbf{Tate}(\mathfrak{A}O_{L_2}, \mathfrak{B}\mathrm{D}^{-1}_{L_2/L_1}), \varepsilon, \frac{dt}{t}\Big)\Big) \\
&= \rho_\beta\Big(a_0 + \sum_{\nu \in (\mathfrak{A}\mathfrak{B}\mathrm{D}^{-1}_{L_2/L_1})^+} a_\nu q^\nu\Big) \\
&= a_0 + \sum_{\delta \in (\mathfrak{A}\mathfrak{B})^+} \left(\sum_{\nu | \mathrm{Tr}_{L_2/L_1}(\nu) = \delta} a_\nu \right) q^\delta.
\end{aligned}$$

17.6 *Compatibilities of U and V operators.* For $i = 1, 2$ let U_i and V_i be the U and V operators on the space of modular forms $\oplus_{\chi\in\mathbb{X}_k(\mathcal{G}_i)} \mathbf{M}_i(k, \mu_N, \chi)$ in characteristic p introduced in 14.7 and 13.1. From the behavior of these operators on weights and q-expansions described in 14.7, in 13.5 and 13.9, we conclude that

$$U_1 \circ \Upsilon^* = \Upsilon^* \circ U_2 \qquad \text{and} \qquad V_1 \circ \Upsilon^* = \Upsilon^* \circ V_2.$$

17.7 Proposition. *(Compatibilities of Θ operators) Let $\mathfrak{P}_1$ be a prime of O_{L_1} over p and let $1 \le i \le f_{\mathfrak{P}_1}$. We have the following identity of differential operators on the algebra $\oplus_\chi \mathbf{M}_2(k, \mu_N, \chi)$:*

$$\Theta_{\mathfrak{P}_1,i} \circ \Upsilon^* = \Upsilon^* \circ \left(\sum_{\mathfrak{P}_2|\mathfrak{P}_1,j|i} e_{\mathfrak{P}_2/\mathfrak{P}_1} \Theta_{\mathfrak{P}_2,j} \right).$$

Here $\sum_{\mathfrak{P}_2|\mathfrak{P}_1,j|i}$ means summing over all primes $\mathfrak{P}_2$ of O_{L_2} over $\mathfrak{P}_1$ and all $1 \le j \le f_{\mathfrak{P}_2}$ such that the embedding $\bar\sigma_{\mathfrak{P}_2,j}: O_{L_2}/\mathfrak{P}_2 \to k$ induces the embedding $\bar\sigma_{\mathfrak{P}_1,i}$ on $O_{L_1}/\mathfrak{P}_1$. As customary, $e_{\mathfrak{P}_2/\mathfrak{P}_1}$ denotes the ramification index of $\mathfrak{P}_2$ relative to $\mathfrak{P}_1$.

Proof: To prove this identity it is enough to show that both sides change the weight and the q-expansion of a modular form $f \in \mathbf{M}_2(k, \mu_N, \chi)$ in the same way.
For weights we argue as follows: the operator $\Upsilon^* \circ \left(\sum_{\mathfrak{P}_2|\mathfrak{P}_1,j|i} e_{\mathfrak{P}_2/\mathfrak{P}_1} \Theta_{\mathfrak{P}_2,j}\right)$ is equal to $\sum_{\mathfrak{P}_2|\mathfrak{P}_1,j|i} e_{\mathfrak{P}_2/\mathfrak{P}_1} \Upsilon^* \circ \Theta_{\mathfrak{P}_2,j}$. Therefore, it is enough to calculate the weight of the modular form $(\Upsilon^* \circ \Theta_{\mathfrak{P}_2,j})(f)$. By 12.38, the modular form $\Theta_{\mathfrak{P}_2,j}(f)$ has weight $\chi \cdot \chi^p_{\mathfrak{P}_2,j-1} \cdot \chi_{\mathfrak{P}_2,j}$. By 17.4, we conclude that $\Upsilon(\Theta_{\mathfrak{P}_2,j}(f))$ has weight $\Psi(\chi) \cdot \chi^p_{\mathfrak{P}_1,i-1} \cdot \chi_{\mathfrak{P}_1,i}$, which is equal to the weight of $(\Theta_{\mathfrak{P}_1,i} \circ \Upsilon^*)(f)$.
We compute the effect on q-expansions. Let $(\mathfrak{A}, \mathfrak{B}, \varepsilon, \mathrm{j})$ be a $\mathfrak{I}_1$-polarized unramified cusp of $\mathfrak{M}_1(k, \mu_N)$. Denote the q-expansion of f at the $\mathfrak{I}_2$-polarized unramified cusp $(\mathfrak{A}O_{L_2}, \mathfrak{B}\mathrm{D}^{-1}_{L_2/L_1}, \varepsilon, \mathrm{j})$ by $a_0 + \sum_\nu a_\nu q^\nu$. By 17.5 and 12.40 the effect of $\Theta_{\mathfrak{P}_1,i}$ on the q-expansion of $\Upsilon^*(f)$ at the cusp $(\mathfrak{A}, \mathfrak{B}, \varepsilon, \mathrm{j})$ is

$$\begin{aligned}
\Theta_{\mathfrak{P}_1,i}(\Upsilon^*(f))(\mathbf{Tate}(\mathfrak{A}, \mathfrak{B}), \varepsilon, \mathrm{j}) &= a_0 + \sum_{\delta\in(\mathfrak{A}\mathfrak{B})^+} \tilde\chi_{\mathfrak{P}_1,i}(\delta) \Bigg(\sum_{\nu|\mathrm{Tr}_{L_2/L_1}(\nu)=\delta} a_\nu \Bigg) q^\delta \\
&= a_0 + \sum_{\delta, \mathrm{Tr}(\nu)=\delta} \Bigg(\sum_{\mathfrak{P}_2|\mathfrak{P}_1,j|i} e_{\mathfrak{P}_2/\mathfrak{P}_1} \tilde\chi_{\mathfrak{P}_2,j}(\nu) a_\nu \Bigg) q^\delta \\
&= \Upsilon^* \Bigg(\sum_{\mathfrak{P}_2|\mathfrak{P}_1,j|i} e_{\mathfrak{P}_2/\mathfrak{P}_1} \Theta_{\mathfrak{P}_2,j}(f) \Bigg).
\end{aligned}$$

17.8 Corollary. *Let Υ^* be the homomorphism from Hilbert modular forms over k of level μ_N w.r.t. O_L to elliptic modular forms over k. Then,*

$$\theta \circ \Upsilon^* = \Upsilon^* \circ \left(\sum_{\mathfrak{P}|p, 1\le i< f_{\mathfrak{P}}} e_{\mathfrak{P}} \Theta_{\mathfrak{P},i} \right),$$

where θ is the classical theta operator of Serre and Swinnerton-Dyer.

18 Integrality and congruences for values of zeta functions.

In this section we apply the results of Section 10 to derive congruences between values of Dedekind zeta functions and bounds on the denominators of these values. In fact, the method is applicable for a wide range of L-functions; [DeRi].

18.1 Definition. *Let*

$$\zeta_L(s) := \sum_I \mathbf{Nm}(I)^{-s} \qquad (\mathrm{Re}(s) > 1)$$

be the Dedekind ζ-function associated to L.

18.2 Theorem. *The function $\zeta_L(s)$ can be continued to a meromorphic function on $\mathbf{C}$, holomorphic for $s \neq 1$. Moreover, $\zeta_L(1-k)$ is in $\mathbf{Q}$ for every integer $k \geq 1$.*

Proof: See [Si].

18.3 Theorem. *Let $k \geq 2$ be an even integer. There exists a $\mathfrak{I}$-polarized modular form*

$$\mathrm{E}_k \in \mathbf{M}\big(\mathbf{C}, \mu_N, \mathbf{Nm}^{k-1}\big)$$

such that the q-expansion of E_k at a $\mathfrak{I}$-polarized unramified cusp $(\mathfrak{A}, \mathfrak{B}, \mathrm{j}_{\mathrm{can}})$, as in 6.4 and 6.11, is

$$\mathbf{Nm}^{k-1}(\mathfrak{A})\Big(2^{-g}\zeta_L(1-k) + \sum_{\nu \in (\mathfrak{A}\mathfrak{B})^+}\Big(\sum_{\nu \in \mathfrak{C} \subset \mathfrak{A}\mathfrak{B}} \mathbf{Nm}(\nu\mathfrak{C}^{-1})^{k-1}\Big)q^\nu\Big).$$

Proof: See [vdG, Chap. I, §6] or [DeRi, Thm. 6.1].

18.4 Definition. *Let $\mathfrak{D}$ be an ideal of O_L dividing p and prime to N. See 3.4 for the notion of $\Gamma_0(p)$-level structures. Let*

$$\pi_{\mathfrak{D}} : \mathfrak{M}\big(\mathbf{C}, \mu_N, \Gamma_0(p)\big) \longrightarrow \mathfrak{M}\big(\mathbf{C}, \mu_N\big)$$

be the map associating to a Hilbert-Blumenthal abelian scheme A over a scheme S with O_L-action, μ_N-level structure and $\Gamma_0(p)$-level structure $H \hookrightarrow A$ the Hilbert-Blumenthal abelian scheme $A/H[\mathfrak{D}]$ over S with induced O_L-action and μ_N-level structure.

18.5 Remark. Here we are forced to work with several polarization modules. Indeed, if the abelian scheme A in 18.4 is $\mathfrak{I}$-polarized, the quotient $A/H[\mathfrak{D}]$ is $\mathfrak{D}\mathfrak{I}$-polarized.

18.6 Theorem. *The notation is as in 18.4. Consider the $\mathfrak{D}\mathfrak{I}$-polarized Eisenstein series*

$$\mathrm{E}_k \in \mathbf{M}\big(\mathbf{C}, \mu_N, \mathbf{Nm}^{k-1}\big).$$

Let

$$H = \left(\frac{O_L}{\mathfrak{D}}\right)(1) \hookrightarrow \mathbf{Tate}\big(O_L, \mathfrak{I}^{-1}\big)_{\sigma_\beta}$$

be the subgroup defined in 6.2. The q-expansion of $\pi_{\mathfrak{D}}^*(\mathrm{E}_k)$ at the $\mathfrak{I}$-polarized cusp $(O_L, \mathfrak{I}^{-1}, \varepsilon, H, \mathrm{j_{can}})$ is

$$\mathbf{Nm}(\mathfrak{D})^{k-1}\Big(2^{-g}\zeta_L(1-k) + \sum_{\nu\in(\mathfrak{I}^{-1}\mathfrak{D})^+}\Big(\sum_{\nu\in\mathfrak{C}\subset\mathfrak{D}\mathfrak{I}^{-1}}\mathbf{Nm}(\nu\mathfrak{C}^{-1})^{k-1}\Big)q^\nu\Big).$$

See 6.7 for the notation.

Proof: The q-expansion of $\pi_{\mathfrak{D}}^*(\mathrm{E}_k)$ is defined by

$$\pi_{\mathfrak{D}}^*(\mathrm{E}_k)\Big(\mathbf{Tate}(O_L,\mathfrak{I}^{-1})_{\sigma_\beta}\otimes_{\mathbf{Z}} R, \varepsilon, H, \frac{dt}{t}\Big);$$

the notation is as in 6.4 and in 6.7. This is equal to

$$\mathrm{E}_k\Big(\pi_{\mathfrak{D}}\big(\mathbf{Tate}(O_L,\mathfrak{I}^{-1})_{\sigma_\beta}\otimes_{\mathbf{Z}} R, \varepsilon, H, \frac{dt}{t}\big)\Big).$$

As explained in 6.3, the abelian scheme $\mathbf{Tate}(O_L,\mathfrak{I}^{-1})_{\sigma_\beta}$ is defined by restricting the semiabelian scheme $\mathrm{D}_L^{-1}\otimes_{\mathbf{Z}}\mathbf{G}_{m,S_{\sigma_\beta}}/\underline{q}(\mathfrak{I}^{-1})$ to the open $S_{\sigma_\beta}\backslash S_{\sigma_\beta,0}$. Using the dictionary of 6.1 we get that $\pi_{\mathfrak{D}}\big(\mathbf{Tate}(O_L,\mathfrak{I}^{-1})_{\sigma_\beta}\big)$ coincides with the restriction to $S_{\sigma_\beta}\backslash S_{\sigma_\beta,0}$ of $\mathfrak{D}^{-1}\mathrm{D}_L^{-1}\otimes_{\mathbf{Z}}\mathbf{G}_{m,S_{\sigma_\beta}}/\underline{q}(\mathfrak{I})$. Observe that $\mathfrak{I}^{-1}\mathfrak{D}\subset\mathfrak{I}^{-1}$, where the inclusion is as O_L-modules of rank 1 with a notion of positivity. Any rational polyhedron $\{\sigma_\beta\}_\beta$ in the given rational polyhedral cone decomposition of the dual cone to $(\mathfrak{I}^{-1})_{\mathbf{R}}^+\subset(\mathfrak{I}^{-1})_{\mathbf{R}}$ induces a rational polyhedron of the dual cone to $(\mathfrak{I}^{-1}\mathfrak{D})_{\mathbf{R}}^+\subset(\mathfrak{I}^{-1}\mathfrak{D})_{\mathbf{R}}$. Hence $\pi_{\mathfrak{D}}\big(\mathbf{Tate}(O_L,\mathfrak{I}^{-1})_{\sigma_\beta}\big)$ is the pullback of $\mathbf{Tate}(\mathfrak{D},\mathfrak{I}^{-1})_{\sigma_\beta}$ via the morphism induced by completing along the boundaries the affine torus embeddings associated to σ_β to the isogeny of the tori

$$(\mathfrak{I}^{-1})^\vee\otimes_{\mathbf{Z}}\mathbf{G}_{m,\mathbf{Z}}\subset(\mathfrak{I}^{-1}\mathfrak{D})^\vee\otimes_{\mathbf{Z}}\mathbf{G}_{m,\mathbf{Z}}.$$

Since the differential dt/t descends to $\mathbf{Tate}(\mathfrak{D},\mathfrak{I}^{-1})$ and corresponds to the differential dt/t on $\mathbf{Tate}(\mathfrak{D},\mathfrak{I}^{-1})$ we conclude.

18.7 Corollary. *There exists a $\mathfrak{I}$-polarized modular form $\mathrm{E}_k^\dagger$ of weight $\mathbf{Nm}^{k-1}$ and level $\mu_N\times\Gamma_0(p)$ i. e., $\mathrm{E}_k^\dagger\in\mathbf{M}(\mathbf{C},\mu_N,\Gamma_0(p),\mathbf{Nm}^{k-1})$, whose q-expansion at the cusp $(O_L,\mathfrak{I}^{-1},\varepsilon,(O_L/p),\mathrm{j_{can}})$ is*

$$\Big(\prod_{\mathfrak{P}|p}(1-\mathbf{Nm}(\mathfrak{P})^{k-1})(2^{-g}\zeta_L(1-k))\Big)+\sum_{\nu\in(\mathfrak{I}^{-1})^+}\Big(\sum_{\mathfrak{C}\subset\mathfrak{I}^{-1}}\mathbf{Nm}'(\nu\mathfrak{C}^{-1})^{k-1}\Big)q^\nu,$$

where

$$\mathbf{Nm}'(\nu\mathfrak{C}^{-1})=\begin{cases}\mathbf{Nm}(\nu\mathfrak{C}^{-1}) & \textit{if } \mathbf{Nm}(\nu\mathfrak{C}^{-1}) \textit{ is prime to } p\\ 0 & \textit{otherwise.}\end{cases}$$

Proof: Let Z be the set of primes of L over p. If I is a subset of Z, define $\mathfrak{D}_I := \prod_{\mathfrak{P}\in I}\mathfrak{P}$. Define

$$\mathrm{E}_k^\dagger := \sum_{I\subset Z}(-1)^{|I|}\pi_{\mathfrak{D}_I}^*(\mathrm{E}_k).$$

Fix $\nu \in (\mathfrak{I}^{-1})^+$. Let I be a subset of Z. It follows from 18.6 that the coefficient of q^ν in the q-expansion at the given cusp of $\pi^*_{\mathfrak{D}_I}(\mathrm{E}_k)$ is 0, if $\nu \notin (\mathfrak{I}^{-1}\mathfrak{D}_I)^+$, and is

$$\mathbf{Nm}^{k-1}(\mathfrak{D}_I)\left(\sum_{\nu\in\mathfrak{C}\subset\mathfrak{I}^{-1}\mathfrak{D}_I}\mathbf{Nm}(\nu\mathfrak{C}^{-1})^{k-1}\right) = \sum_{\nu\in\mathfrak{C}\subset\mathfrak{I}^{-1}\mathfrak{D}_I}\mathbf{Nm}\left(\nu\left(\mathfrak{C}\mathfrak{D}_I^{-1}\right)^{-1}\right)^{k-1}$$
$$= \sum_{\mathfrak{W}\subset\mathfrak{I}^{-1}|\nu\in\mathfrak{W}\mathfrak{D}_I}\mathbf{Nm}\left(\nu\mathfrak{W}^{-1}\right)^{k-1}$$

otherwise. Fix $\mathfrak{W} \subset \mathfrak{I}^{-1}$ such that $\nu \in \mathfrak{W}$. Let I be the maximal subset of Z such that $\mathfrak{D}_I|(\nu\mathfrak{W}^{-1})$. The contribution of $\mathbf{Nm}(\nu\mathfrak{W}^{-1})$ to the coefficient of q^ν in the q-expansion of $\mathrm{E}_k^\dagger$ is

$$\left(\sum_{I'\subset I}(-1)^{|I'|}\right)\mathbf{Nm}(\nu\mathfrak{W}^{-1}).$$

Since $\sum_{I'\subset I}(-1)^{|I'|}$ is 0 if $I \neq \emptyset$ and is 1 if $I = \emptyset$, we conclude.

18.8 *p-adic Eisenstein series.* Let p be a prime not dividing N. Let $k_1 < k_2 < \ldots < k_n < \ldots$ be a sequence of even integers ≥ 2 converging p-adically to $k \in \mathbf{Z}_p$. Let $(O_L, \mathfrak{I}^{-1}, \mathrm{j})$ be a $\mathfrak{I}$-polarized unramified cusp defined over $\mathbf{Z}_{(p)}$. It follows from 18.3 that for every $\nu \in (\mathfrak{I})^+$ the coefficient $a_{k_i,\nu}$ of q^ν in the q-expansion of E_{k_i} at the given cusp is

$$a_{k_i,\nu} = \left(\sum_{\nu\in\mathfrak{C}\subset\mathfrak{I}^{-1}}\mathbf{Nm}(\nu\mathfrak{C}^{-1})^{k_i-1}\right) \in \mathbf{Q}.$$

It follows from the q-expansion principle, see 6.10, that

$$E_{k_i} \in \mathbf{M}\big(\mathbf{Q}, \mu_N, \mathbf{Nm}^{k_i-1}\big).$$

For $i \longrightarrow \infty$ we have

$$\lim_{i\to\infty} a_{k_i,\nu} = \Big(\sum_{\nu\in\mathfrak{C}\subset\mathfrak{I}^{-1}}\mathbf{Nm}'(\nu\mathfrak{C}^{-1})^{k-1}\Big)$$

and the convergence is uniform in k_i. See 18.7 for the definition of $\mathbf{Nm}'$. It follows from 10.7 that the sequence

$$a_{k_i,0} = 2^{-g}\,\zeta_L(1-k_i) \in \mathbf{Q}_p$$

is bounded. As in [Se, Cor. 2, §1.5], one concludes that it converges p-adically. Define

$$\zeta_L^*(1-k) := \lim_{i\to\infty}\zeta_L(1-k_i) \in \mathbf{Q}_p.$$

One may interpret this formula as the value of the p-adic zeta function ζ_L^* associated to L at $1-k$; see [Se, Thm. 3, §1.6, Thm. 20, §5.3]. We also get that

$$\mathrm{E}_k^* := \frac{\zeta_L^*(1-k)}{2^g} + \sum_{\nu\in\in\mathfrak{I}^+}\Big(\sum_{\nu\mathfrak{C}\subset\mathfrak{I}}\mathbf{Nm}'(\nu\mathfrak{C}^{-1})^{k-1}\Big)q^\nu$$

is a p-adic modular form à la Serre; see 10.8.

18.9 $\mathrm{E}_k^\dagger$ *as a p-adic Eisenstein series.* Let k be an even, positive integer. By 11.13 the $\mathfrak{I}$-polarized modular form $\mathrm{E}_k^\dagger$ defines a $\mathfrak{I}$-polarized p-adic modular form à la

Katz, and hence à la Serre by 11.11, of level μ_N over $\mathbf{Q}_p$. It has the property that its q-expansion at a $\mathfrak{J}$-polarized cusp $(\mathfrak{A}, \mathfrak{B}, \varepsilon_{p^\infty N}, \mathrm{j}_\varepsilon)$, in the sense of 10.10, is the q-expansion of $\mathrm{E}_k^\dagger$ at the $\mathfrak{J}$-polarized cusp $(\mathfrak{A}, \mathfrak{B}, \varepsilon_N, H, \mathrm{j}_\varepsilon)$. The latter is given in 18.7. In particular, it has the same coefficients for $\nu \neq 0$ as the q-expansion of the Serre $\mathfrak{J}$-polarized p-adic Hilbert modular form E_k^* of level μ_N at the cusp $(\mathfrak{A}, \mathfrak{B}, \varepsilon, \mathrm{j}_\varepsilon)$. It follows from 10.11, more precisely from the generalization of 10.7 to the case of p-adic modular forms of level μ_N, that $\mathrm{E}_k^\dagger$ has *the same q-expansion* as E_k^*. Hence,

$$\zeta_L^*(1-k) = \prod_{\mathfrak{P}|p} \left((1 - \mathbf{Nm}(\mathfrak{P})^{k-1})\right) \cdot \zeta_L(1-k) \qquad \text{for } k \in 2\mathbf{Z},\ k \geq 2.$$

Compare with [Se, Rmk. 1, §1.6].

18.10 *Notation.* Let $\mathfrak{P}$ be a prime dividing p in O_L. Let $B_{\mathfrak{P}}$ be the maximal abelian subextension over $\mathbf{Q}_p$ of the completion $L_{\mathfrak{P}}$ of L at $\mathfrak{P}$ and let $A_{\mathfrak{P}}$ be its ring of integers. Let $e'(\mathfrak{P}/p)$ be the ramification index of $B_{\mathfrak{P}}$ over $\mathbf{Q}_p$. Note that $e'(\mathfrak{P}/p)$ divides $e_{\mathfrak{P}}$. Write $e'(\mathfrak{P}/p) = e'(\mathfrak{P}/p)^{\mathrm{t}} \cdot e'(\mathfrak{P}/p)^{\mathrm{w}}$, where $e'(\mathfrak{P}/p)^{\mathrm{t}}$ is the prime to p part of $e'(\mathfrak{P}/p)$. Local class field theory gives that

$$H(\mathfrak{P}) := \mathbf{Nm}_{B_{\mathfrak{P}}/\mathbf{Q}_p}(A_{\mathfrak{P}}^*)$$

is equal to $\mathbf{Nm}_{L_{\mathfrak{P}}/\mathbf{Q}_p}(O_{L_{\mathfrak{P}}}^*)$. Moreover, for $p \neq 2$,

$$H(\mathfrak{P}) = H_1(\mathfrak{P})\, H_2(\mathfrak{P}),$$

where $H_1(\mathfrak{P}) \subset \mu_{p-1}$ is the unique subgroup of μ_{p-1} of index $e'(\mathfrak{P}/p)^{\mathrm{t}}$ and $H_2(\mathfrak{P}) \subset 1 + p\mathbf{Z}_p$ is equal to $1 + pe'(\mathfrak{P}/p)^{\mathrm{w}}\mathbf{Z}_p$, is the unique subgroup of $1 + p\mathbf{Z}_p$ of index $e'(\mathfrak{P}/p)^{\mathrm{w}}$. For $p = 2$, $H(\mathfrak{P})$ is the unique subgroup of index $e'(\mathfrak{P}/p)$ of $\mathbf{Z}_2^* \cong \{\pm 1\} \times \mathbf{Z}_2$.

Let

$$e_p^{\mathrm{t}} := \min\{e'(\mathfrak{P}/p)^{\mathrm{t}} : \mathfrak{P}|p\}, \qquad e_p^{\mathrm{w}} := \min\{e'(\mathfrak{P}/p)^{\mathrm{w}} : \mathfrak{P}|p\}.$$

For $p = 2$ we use

$$e_2 := e_2^{\mathrm{w}}.$$

We find that for $p \neq 2$,

$$H := \mathbf{Nm}\left((O_L \otimes \mathbf{Z}_p)^*\right) = \prod_{\mathfrak{P}|p} H_1(\mathfrak{P}) \times \prod_{\mathfrak{P}|p} H_2(\mathfrak{P}) = H_1 \times H_2,$$

where H_1 is the unique subgroup of index e_p^{t} of μ_{p-1} and H_2 is the unique subgroup of e_p^{w} of $1 + p\mathbf{Z}_p$.

For $p = 2$, we find that

$$H := \mathbf{Nm}\left((O_L \otimes \mathbf{Z}_p)^*\right) = \prod_{\mathfrak{P}|p} H(\mathfrak{P})$$

is a subgroup of index e_2 of $\mathbf{Z}_2^*$.

18.11 *Calculation of exponents.* Let $n \geq 1$ be an integer. We compute the exponent of the abelian group

$$\overline{H} := \mathrm{Image}(H) \quad \text{via the map } \mathbf{Z}_p^* \to \left(\mathbf{Z}_p/p^n\mathbf{Z}_p\right)^*.$$

If $x \in \mathbf{R}$, we use the notation

$$\lceil x \rceil := \min \{n \in \mathbf{Z} | x \leq n\}.$$

If $p \neq 2$, the exponent is equal to

$$\frac{p-1}{e_p^{\mathrm{t}}} \cdot \left\lceil \frac{p^{n-1}}{e_p^{\mathrm{w}}} \right\rceil .$$

If $p = 2$, the exponent is equal to

$$2^{\epsilon} \cdot \left\lceil \frac{2^{\mathfrak{l}(n)}}{e_2} \right\rceil ,$$

where $\epsilon = 0, 1$ depending on the case and

$$\mathfrak{l}(n) := \begin{cases} n-1 & \text{if } n \leq 2, \\ n-2 & \text{if } n \geq 3; \end{cases}$$

(note that $2^{\mathfrak{l}(n)}$ is the exponent of $\left(\mathbf{Z}_2/2^n\mathbf{Z}_2\right)^*$).

18.12 Theorem. *Let $k > 1$ be an even integer. Suppose that $2^{-g}\zeta_L(1-k)$ is not p-integral and let $n = -\mathrm{val}_p\left(2^{-g}\zeta_L(1-k)\right)$. Then,*

i. *if $p \neq 2$,*

$$k \equiv 0 \mod \frac{p-1}{e_p^{\mathrm{t}}} \cdot \left\lceil \frac{p^{n-1}}{e_p^{\mathrm{w}}} \right\rceil ;$$

ii. *if $p = 2$,*

$$k \equiv 0 \mod \left\lceil \frac{2^{\mathfrak{l}(n)}}{e_2} \right\rceil .$$

Proof: Consider the modular form $f_1 := p^n \mathrm{E}_k$. It is a modular form of weight $\mathbf{Nm}^k$ over $\mathbf{Z}_p$ and the reduction modulo p^n of its q-expansion is equivalent to $2^{-g}p^n\zeta_L(1-k)$. Let f_2 be the modular form $2^{-g}p^n\zeta_L(1-k)$ over $\mathbf{Z}_p$ of weight 0. By 10.5 we conclude that $\mathbf{Nm}^k \in \mathbb{X}_{\mathbf{Z}/p^n\mathbf{Z}}(n)$ i. e., that for any $b \in \left(O_L \otimes_{\mathbf{Z}} \mathbf{Z}_p\right)^*$ we have $\mathbf{Nm}^k(b) \equiv 1$ modulo p^n. It follows that the exponent of $\overline{H}$ (see 18.11) divides k and the theorem follows.

18.13 Theorem. *Let $k, k' \geq 2$ be even integers such that $k \equiv k'$ modulo $(p-1)p^m$ for some non-negative integer m. Then*

i. *if $k \not\equiv 0$ mod $(p-1)/e_p^{\mathrm{t}}$ (and hence $p \neq 2$), then*

$$\mathrm{val}_p \left\{ \left(\prod_{\mathfrak{P}|p} (1 - \mathbf{Nm}(\mathfrak{P})^{k-1}) \right) \frac{\zeta_L(1-k)}{2^g} - \left(\prod_{\mathfrak{P}|p} (1 - \mathbf{Nm}(\mathfrak{P})^{k'-1}) \right) \frac{\zeta_L(1-k')}{2^g} \right\}$$
$$\geq m + 1;$$

ii. *if $k \equiv 0$ mod $(p-1)/e_p^{\mathrm{t}}$ and $p \neq 2$, then*

$$\mathrm{val}_p \left\{ \left(\prod_{\mathfrak{P}|p} (1 - \mathbf{Nm}(\mathfrak{P})^{k-1}) \right) \frac{\zeta_L(1-k)}{2^g} - \left(\prod_{\mathfrak{P}|p} (1 - \mathbf{Nm}(\mathfrak{P})^{k'-1}) \right) \frac{\zeta_L(1-k')}{2^g} \right\}$$
$$\geq m - \mathrm{val}_p(kk') - 1 - 2\mathrm{val}_p\left(e_p^{\mathrm{w}}\right);$$

iii. *if $p = 2$, then*

$$\mathrm{val}_2\left\{\left(\prod_{\mathfrak{P}|p}(1-\mathbf{Nm}(\mathfrak{P})^{k-1})\right)\frac{\zeta_L(1-k)}{2^g} - \left(\prod_{\mathfrak{P}|p}(1-\mathbf{Nm}(\mathfrak{P})^{k'-1})\right)\frac{\zeta_L(1-k')}{2^g}\right\}$$

$$\geq m - 2 - \mathrm{val}_2(kk') - 2\mathrm{val}_2(e_2).$$

Proof: Let

$$\ell := \max\left\{-\mathrm{val}_p(2^{-g}\zeta_L(1-k)), -\mathrm{val}_p(2^{-g}\zeta_L(1-k')), 0\right\}$$

and let

$$\beta = p^\ell\left(2^{-g}\zeta_L(1-k)\prod_{\mathfrak{P}|p}(1-\mathbf{Nm}(\mathfrak{P})^{k-1}) - 2^{-g}\zeta_L(1-k')\prod_{\mathfrak{P}|p}(1-\mathbf{Nm}(\mathfrak{P})^{k'-1})\right).$$

Let $i := 1$ if $p \neq 2$ and let $i := 2$ if $p = 2$. Note that if x is an integer prime to p, then

$$x^k - x^{k'} \equiv 0 \mod p^{m+i}.$$

It follows that

$$f := p^\ell \mathrm{E}_k^\dagger - p^\ell \mathrm{E}_{k'}^\dagger - \beta \equiv 0 \qquad \mathrm{mod} \quad p^{m+i+\ell}.$$

Using 18.9, we interpret f as a p-adic modular form à la Katz. It reduces to function 0 on $\mathfrak{M}(\mathbf{Z}/p^{m+i+\ell}, \mu_{Np^{m+i+\ell}})$, invariant under $\Gamma_{m+i+\ell}$. (Here N is any auxiliary integer ≥ 4 and prime to p). It follows that for all $\alpha \in \Gamma_{m+i+\ell} = (O_L/p^{m+i+\ell})^*$ we have

$$\alpha^* f - f = (\mathbf{Nm}^k(\alpha) - 1)p^\ell E_k^\dagger - (\mathbf{Nm}^{k'}(\alpha) - 1)p^\ell E_{k'}^\dagger \equiv 0 \qquad \mathrm{mod} \quad p^{m+i+\ell}.$$

Consider this equation modulo p^{m+i}. For $\alpha \in \Gamma_{m+i+\ell}$ we have $\mathbf{Nm}^k(\alpha) \equiv \mathbf{Nm}^{k'}(\alpha)$ modulo p^{m+i}. Thus, we find

$$\alpha^* f - f = (\mathbf{Nm}^k(\alpha) - 1)\left(p^\ell E_k^\dagger - p^\ell E_{k'}^\dagger\right) \equiv 0 \qquad \mathrm{mod} \quad p^{m+i},$$

and from here that

$$(\mathbf{Nm}^k(\alpha) - 1)\beta \equiv 0 \qquad \mathrm{mod} \quad p^{m+i}.$$

Let t be the p-adic valuation of β. Then $\mathbf{Nm}^k(\alpha) - 1 \equiv 0$ modulo $\lceil p^{m+i-t} \rceil$. Let $n := m + i - t$. In the notation of 18.10 and 18.11, $H = \mathbf{Nm}((O_L \otimes \mathbf{Z}_p)^*)$ and we have:

1) If $p \neq 2$, then $H = H_1 \times H_2$ where H_1, H_2 are as in loc. cit.

1.a) If $k \not\equiv 0 \bmod \frac{p-1}{e_p^{\mathrm{t}}}$, then we must have $t \geq m + i$. In this case also $\ell = 0$ by 18.12. Part (i) follows.

1.b) If $k \equiv 0 \bmod \frac{p-1}{e_p^{\mathrm{t}}}$, then 18.10 implies that $k \equiv 0 \bmod \left\lceil \frac{p^{n-1}}{e_p^{\mathrm{w}}} \right\rceil$. Therefore, $n - 1 - \mathrm{val}_p(e_p^{\mathrm{w}}) \leq \mathrm{val}_p(k)$ and we get that $t \geq m + i - 1 - \mathrm{val}_p(e_p^{\mathrm{w}}) - \mathrm{val}_p(k)$. The same holds for k' and we conclude that

$$\mathrm{val}_p(p^{-\ell}\beta) \geq m + i - 1 - \mathrm{val}_p(e_p^{\mathrm{w}}) - \ell - \min\{\mathrm{val}_p(k), \mathrm{val}_p(k')\}.$$

However, by 18.12, $\ell \leq \max\{\mathrm{val}_p(k), \mathrm{val}_p(k')\} + \mathrm{val}_p(e_p^{\mathrm{w}}) + 1$. Put together this yields

$$\mathrm{val}_p(p^{-\ell}\beta) \geq m - 1 - 2\mathrm{val}_p(e_p^{\mathrm{w}}) - \mathrm{val}_p(kk').$$

This proves Part (ii).

2) If $p = 2$, H is a subgroup of index e_2 of $\mathbf{Z}_2^*$. By 18.11 its image $\overline{H}$ in $\left(\mathbf{Z}_2^*/2^n\mathbf{Z}_2\right)^*$ is killed by $\left\lceil \frac{2^{\mathfrak{l}(n)}}{e_2} \right\rceil$, but not by any smaller power of 2. Thus, $\mathrm{val}_2(k) \geq \mathfrak{l}(n) - \mathrm{val}_2(e_2) = \mathfrak{l}(m+2-t) - \mathrm{val}_2(e_2)$. Recall that $t \geq 0$ and $m \geq 1$. Thus, $t \geq m - \mathrm{val}_2(e_2) - \mathrm{val}_2(k)$. Therefore,

$$\mathrm{val}_2\left(2^{-\ell}\beta\right) \geq m - \mathrm{val}_2(e_2) - \ell - \min\left\{\mathrm{val}_2(k), \mathrm{val}_2(k')\right\}.$$

Using 18.12, we find that $\mathfrak{l}(\ell) - \mathrm{val}_2(e_2) \leq \mathrm{val}_2(k)$ and therefore,

$$\ell \leq 2 + \max\left\{\mathrm{val}_2(k), \mathrm{val}_2(k')\right\} + \mathrm{val}_2(e_2).$$

Part (iii) follows from the last two displayed formulas.

18.14 Remark. For $p = 2$, one can improve the results given the precise structure of the completions of L at primes above 2 e. g., when 2 is inert; c.f. [Go2].

19 Numerical examples.

19.1 *Example 1.* Consider the field $L = \mathbf{Q}(\sqrt{3})$. It is a real quadratic field of discriminant 12, equal to the totally real subfield of the cyclotomic field obtained by adjoining to $\mathbf{Q}$ the roots of unity of order 12. The following table provides some information on the decomposition of rational primes in L.

p	decomposition	e_p^{tame}	e_p^{Wild}
2	ramified	1	2
3	ramified	2	1
$5, 7, 17, 19, 29, 31$	inert	1	1
$11, 13, 23, 37$	split	1	1

The results of the Section 11 imply that the only odd primes at which $\zeta_L(1-k)^{-1}$ can have positive p-adic valuation n are the primes p such that $(p-1)|k$, and then $n-1$ is at most the power of p dividing k. For the prime $p = 2$, we find that if n is the valuation at 2 of $2^g\zeta_L(1-k)^{-1}$, then $\mathrm{val}_2(k) \geq \mathfrak{l}(n) - 1$. In this case, the ϵ in 18.11 is 1 since -1 is not a norm from $\mathbf{Q}[\sqrt{3}]$. Hence, the bound may be improved to $\mathrm{val}_2(k) \geq \mathfrak{l}(n)$. For example, taking $k = 18$ the prediction is that the only odd primes at which $\zeta_L(-17)^{-1}$ may have positive valuation n are $3, 7$ and 19 and that valuation can be at most $3, 1$ and 1, respectively. At 2 the valuation can be at most 1. Indeed:

$$\begin{aligned}\zeta_L(-17) =& 514802473837215246476827/7182 \\ =& 2^{-1} \cdot 3^{-3} \cdot 7^{-1} \cdot 11 \cdot 19^{-1} \cdot 43867 \cdot 1066866320794499171.\end{aligned}$$

Another interesting value is the denominator of $\zeta_L(-35)$, which is

$$\mathrm{denominator}\left(\zeta_L(-35)\right) = 2^2 \cdot 3^3 \cdot 5 \cdot 7 \cdot 13 \cdot 19 \cdot 37.$$

We consider the congruences involving $2^{-2}\zeta_L(1-2)$ and $2^{-2}\zeta_L(1-26)$. Congruences are predicted for the primes $2, 3, 5, 7, 13$. The prediction is

$$\mathrm{val}_2\left\{(1-2^{2-1})2^{-2}\zeta_L(1-2) - (1-2^{26-1})2^{-2}\zeta_L(1-26)\right\} \geq 3-2-2-2 = -3.$$

Using $\zeta_L(-1) = 1/6$ and

$$\zeta_L(-25) = 596034262439124086786635474736705480 11/6$$

one verifies the congruence since the valuation is -1.
For the prime 3 the prediction is

$$\mathrm{val}_3\left\{(1-3)2^{-2}\zeta_L(-1) - (1-3^{25})2^{-2}\zeta_L(-25)\right\} \geq 1 - 0 - 1 - 2\cdot 0 = 0.$$

This indeed holds, since the valuation is 0. For the prime 7, we expect

$$\mathrm{val}_7\left\{(1-49)2^{-2}\zeta_L(-1) - (1-49^{25})2^{-2}\zeta(-25)\right\} \geq 0 + 1 = 1,$$

which holds, since the valuation is 1. For the prime 13 the predicted congruence is

$$\mathrm{val}_{13}\left\{(1-13)^2 2^{-2}\zeta_L(-1) - (1-13^{25})^2 2^{-2}\zeta_L(-25)\right\} \geq 0 + 1 = 1.$$

This is verified, since the valuation is 1.

19.2 *Example 2.* Consider the cyclic cubic totally real field L of discriminant 49, equal to the totally real subfield of the cyclotomic field of roots of unity of order 7. The following table provides some information on the decomposition of rational primes in L.

p	decomposition	e_p^{tame}	e_p^{Wild}
7	ramified	3	1
2, 3, 5, 11, 17, 19, 23, 31, 37	inert	1	1
13, 29	split	1	1

The results of Section 11 imply that if p is odd, not equal to 7, then the only odd primes at which $\zeta_L(1-k)^{-1}$ can have positive p-adic valuation n are the primes p such that $(p-1)|k$ and then $n-1$ is at most the power of p dividing k. If $p = 7$, then $\zeta_L(1-k)^{-1}$ can indeed have positive 7-adic valuation n (k is even) and then $n-1$ is at most the power of 7 dividing k. For the prime $p = 2$, letting $n := \mathrm{val}_2(2^3\zeta_L(1-k)^{-1})$, we find that $\mathrm{val}_2(k) \geq \mathfrak{l}(n)$. This implies that $\mathrm{val}_2(\mathrm{denom}.\zeta_L(1-k)) \leq \mathrm{val}_2(k) - 1$. For example, taking $k = 10$ the prediction is that the only odd primes at which $\zeta_L(-9)^{-1}$ may have positive valuation n are 3, 7 and 11, and that valuation can be at most 1 in each case. At 2, the valuation cannot be positive. Indeed:

$$\zeta_L(-9) = -1141452324871/231 = -3^{-1}\cdot 7^{-1}\cdot 11^{-1}\cdot 1141452324871.$$

We consider congruence for 7 for $2^{-3}\zeta_L(1-2)$ and $2^{-3}\zeta_L(1-14)$. The expected congruence is

$$\mathrm{val}_7\left\{(1-7^{3(2-1)})2^{-3}\zeta_L(1-2) - (1-7^{3(16-1)})2^{-3}\zeta_L(1-14)\right\} \geq -1 - 1 = -2.$$

It holds because the denominators of both values,

$$\zeta_L(-1) = -1/21, \zeta_L(-13) = -55890871330157828667 37/147$$

are not divisible by 7^3. For 2 the expected congruence is

$$\mathrm{val}_2\left\{(1-8)2^{-3}\zeta_L(-1) - (1-8^{15})2^{-3}\zeta_L(-13)\right\} \geq 2 - 2 - 2 - 2\cdot 0 = -2.$$

This is visibly true, because both zeta values have odd numerator. For $p = 13$, the expected congruence is

$$\mathrm{val}_{13}\left\{(1-13)^3 2^{-3}\zeta_L(-1) - (1-13^{25})2^{-3}\zeta_L(-13)\right\} \geq 0+1 = 1.$$

This holds, since the valuation is 1.

19.3 *Example 3.* Take the non-Galois totally real cubic field $L = \mathbf{Q}[x]/(x^3-9x-6)$. It has discriminant $2^3 \cdot 3^5$. The prime 2 decomposes as $\mathfrak{P}_1^2\mathfrak{P}_2$ and therefore $e_2 = 1$. The prime 3 decomposes as $\mathfrak{P}^3$, the field L_3 is cubic non-Galois and, therefore, $e_3^{\mathrm{t}} = e_3^{\mathrm{w}} = 1$. We conclude that if an odd prime divides the denominator of $\zeta_L(1-k)$ with valuation n, then $k \equiv 0$ modulo $(p-1)p^{n-1}$. Analogously, if n is the valuation at 2 of $2^3\zeta_L(1-k)^{-1}$, then $\mathrm{val}_2(k) \geq \mathfrak{l}(n)$. For example, if $k = 6$, we find

$$\zeta_L(-5) = -2 \cdot 3^{-2} \cdot 5^2 \cdot 7^{-1} \cdot 184669 \cdot 512249.$$

The prime 7 decomposes as a product of two prime ideals in L and $e_7^{\mathrm{t}} = e_7^{\mathrm{w}} = 1$. The expected congruence for $2^{-3}\zeta_L(1-2)$ and $2^{-3}\zeta_L(1-14)$ is

$$(1-7)(1-7^2)2^{-3}\zeta_L(-1) \equiv (1-7^{15})(1-7^{30})2^{-3}\zeta_L(-13) \pmod{7}.$$

Using that $\zeta_L(-1) = -70/3$ and

$$\zeta_L(-13) = -43346131550431228090356336024418702874761 0/3$$

are both divisible by 7, the congruence follows trivially. For the values $2^{-3}\zeta_L(1-4)$ and $2^{-3}\zeta_L(1-16)$ we again predict

$$\mathrm{val}_7\left\{(1-7^3)(1-7^6)2^{-3}\zeta_L(-3) - (1-7^{15})(1-7^{30})2^{-3}\zeta_L(-15)\right\} \geq 1.$$

The zeta values are $\zeta(-3) = 2556221/15$ and

$$\zeta_L(-15) = 838225008486241735965907905513225151275805634985499 57/1020,$$

which are both 7-adic units. The congruence holds, though, since the valuation is 1. For the prime 2, we predict that

$$\mathrm{val}_2\left\{(1-2^3)(1-4^3)2^{-3}\zeta_L(-3) - (1-2^{15})(1-4^{15})\zeta_L(-15)\right\} \geq 2-2-6 = -6.$$

This holds, since the valuation is -5. Again for the prime 2 we predict

$$\mathrm{val}_2\left\{(1-2)^2 2^{-3}\zeta_L(-1) - (1-2^{17})^2 2^{-3}\zeta_L(-17)\right\} \geq 4-2-2-2\cdot 0 = 0.$$

Using

$$\zeta_L(-17) = -\big(3647421225841578953319613809666454838832065018732125543 \\ 06326430\big) \cdot 3591^{-1}$$

one verifies that the valuation is 1. In particular, the congruence holds.

20 Comments regarding values of zeta functions.

We make a few remarks on values of zeta functions. We have no real theorem to offer here, but rather we would like to point out some facts concerning Bernoulli numbers that are interesting in the context of this manuscript. The connection to zeta functions rests on the identity

$$\zeta(1-2k) = -\frac{B_{2k}}{2k}, \quad k \geq 1.$$

To begin with, consider the question of which normalized Eisenstein series E_{2k} for the modular group $\mathrm{SL}_2(\mathbf{Z})$ are congruent to one modulo p. Here p is a fixed odd prime. Since the Eisenstein series has the form

$$1+\frac{2}{\zeta(1-2k)}\sum_{n=1}^{\infty}\sigma_{2k-1}(n)q^n,$$

the question is for which k does p divide the denominator of $\zeta(1-2k)$? (As an aside we mention that this means that there is an element of order p is the $4k-1$ stable homotopy groups of the spheres. Cf. [MiSt, App. B]; note that B_n in their notation is our B_{2n}). The Kummer congruences imply that this is the case iff $2k \equiv 0 \pmod{p-1}$ and then

$$\mathrm{val}_p(\zeta(1-2k)) = -1-\mathrm{val}_p(2k);$$

see [Se, §1.1]. Assume, for argument's sake, that k is prime. Then $p-1$ is either 1, 2, k or $2k$. Apart from $p=2,3$, assuming $k>3$, we are left only with the possibility that $p=2k+1$, i. e., that k is a Germain prime! There are infinitely many primes that are *not* Germain primes (take k to be a prime congruent to 1 mod 6), hence the order of the denominator of $\zeta(1-2k)$ does not grow to infinity with k. In fact, one can prove that if all the prime factors of k are congruent to 1 modulo 6 then the denominator of $\zeta(1-2k)$ is equal to 12 ([MiSt, App. B, Pb. B-1]).

Let $p \equiv 1 \pmod 4$ and let $K=\mathbf{Q}(\sqrt{p})$. Let $\chi = \left(\frac{\cdot}{p}\right)$ be the corresponding Dirichlet character. Let ω be the Teichmüller character, then $\chi = \omega^{(p-1)/2}$. We note also that the discriminant of L is p and is equal to the conductor f_χ of χ.

20.1 Proposition. *Let p be a prime congruent to 1 modulo 4 and let $K := \mathbf{Q}(\sqrt{p})$. Assume Vandiver's conjecture ([Wa, p. 159]). Then for any integer $r \geq 1$ we have*

$$\mathrm{val}_p(\zeta_K(1-r(p-1))) = -1-\mathrm{val}_p(r).$$

Hence, in this case, the Hasse invariant in characteristic p lifts to the Eisenstein series $\mathrm{E}_{K,p-1}$.

Proof: We have the identity

$$\zeta_K(1-r(p-1)) = \zeta_{\mathbf{Q}}(1-r(p-1))\cdot L(1-r(p-1),\chi).$$

Therefore, the claim would follow if we prove that $\mathrm{val}_p(L(1-r(p-1),\chi))=0$. Using [Wa, Thms 4.2, 5.11] we get that for any $n\geq 1$

$$L_p(1-n,\chi) = (1-\chi\omega^{-n}(p)p^{n-1})\cdot L(1-n,\chi\omega^{-n}).$$

Applying this formula in our situation we get

$$\mathrm{val}_p(L(1-r(p-1),\chi)) = \mathrm{val}_p(L_p(1-r(p-1),\chi)).$$

By [Wa, Cor. 5.13], for any integers m,n we have $L_p(m,\chi)\equiv L_p(n,\chi)$ mod. p and both numbers are p-integral. It is thus enough to prove for a *single* integer m that $L_p(m,\chi)$ is a p-adic unit. Using the results cited above we find

$$\begin{aligned}\mathrm{val}_p(L_p(1-(p-1)/2,\chi)) &= \mathrm{val}_p(L(1-(p-1)/2,\chi\omega^{(p-1)/2}))\\ &= \mathrm{val}_p(\zeta(1-(p-1)/2))\\ &= \mathrm{val}_p(B_{(p-1)/2}).\end{aligned}$$

The assertion $\mathrm{val}_p(B_{(p-1)/2}) = 0$ is known as the Ankeny-Artin-Chowla conjecture and was verified for $p < 10^{11}$ by A. J. van der Poorten, H. te Riele and H. Williams in [vdP]. This conjecture is a consequence of Vandiver's conjecture. See [Wa, Thm. 5.34].

20.2 Proposition. *Let p be a prime congruent to 1 modulo 4 and let $K = \mathbf{Q}(\sqrt{p})$. For every odd integer $r \geq 1$*

$$\mathrm{val}_p(\zeta_K(1 - r(p-1)/2)) = -1 - \mathrm{val}_p(r).$$

Hence the partial Hasse invariant defined in 7.12 admits a lift to char. 0, the Eisenstein series $\mathrm{E}_{K,(p-1)/2}$.

Proof: We have

$$\zeta_K(1 - r(p-1)/2) = \zeta_{\mathbf{Q}}(1 - r(p-1)/2) \cdot L(1 - r(p-1)/2, \chi).$$

Note that $r(p-1)/2 \not\equiv 0 \bmod p-1$ and therefore

$$\mathrm{val}_p(\zeta_K(1-r(p-1)/2)) = -1-\mathrm{val}_p(r) \Leftrightarrow \mathrm{val}_p(L(1-r(p-1)/2,\chi)) = -1-\mathrm{val}_p(r).$$

However,

$$\begin{aligned} L_p(1 - r(p-1)/2, \mathbf{1}) &= (1 - \omega^{r(p-1)/2}(p)p^{r(p-1)/2-1})L(1 - r(p-1)/2, \omega^{r(p-1)/2}) \\ &= L(1 - r(p-1)/2, \chi), \end{aligned}$$

and therefore,

$$\mathrm{val}_p(L(1 - r(p-1)/2, \chi)) = \mathrm{val}_p(L_p(1 - r(p-1)/2, \mathbf{1})).$$

Using [Wa, Ex. 5.11], we write

$$L_p(s, \mathbf{1}) = \frac{p-1}{p} \cdot \frac{1}{s-1} + a_0 + a_1(s-1) + a_2(s-1)^2 + \dots,$$

where $a_i \in \mathbf{Z}_p$ for every i. In particular, $\mathrm{val}_p(L_p(1 - r(p-1)/2, \mathbf{1})) = -1 - \mathrm{val}_p(r)$.

21 References.

[AG] Andreatta, F., Goren, E. Z.: Hilbert modular varieties of low dimension. To appear in *Geometric Aspects of Dwork's Theory, A Volume in memory of Bernard Dwork*, Alan Adolphson, Francesco Baldassarri, Pierre Berthelot, Nicholas Katz, and Francois Loeser (Eds.), 62 pages.

[Ch] Chai, C.-L.: Arithmetic minimal compactification of the Hilbert-Blumenthal moduli spaces. Appendix to "The Iwasawa conjecture for totally real fields" by A. Wiles. *Ann. of Math.* (2) 131 (1990), no. 3, 541–554.

[DePa] Deligne, P., Pappas, G.: Singularités des espaces de modules de Hilbert, en les caractéristiques divisant le discriminant. *Compositio Math.* 90 (1994), no. 1, 59–79.

[DeRi] Deligne, P., Ribet, K. A.: Values of abelian L-functions at negative integers over totally real fields. *Invent. Math.* 59 (1980), no. 3, 227–286.

[vdG] Van der Geer, G.: *Hilbert modular surfaces.* Ergebnisse der Mathematik und ihrer Grenzgebiete (3), 16, Springer-Verlag, 1988.

[Go1] Goren, E. Z.: Hasse invariants for Hilbert modular varieties. *Israel J. Math.*, 122 (2001), 157–174.

[Go2] Goren, E. Z.: Hilbert modular forms modulo p^m: the unramified case. *J. Number Theory* 90 (2001), no. 2, 341–375.

[Go3] Goren, E. Z.: *Lectures on Hilbert modular varieties and modular forms.* With the assistance of Marc-Hubert Nicole. CRM Monograph Series, 14. American Mathematical Society, Providence, RI, 2002.

[GoOo] Goren, E. Z., Oort, F.: Stratifications of Hilbert modular varieties, *J. Algebraic Geometry* 9 (2000), 111-154.

[Gr] Gross, B. H.: A tameness criterion for Galois representations associated to modular forms (mod p). *Duke Math. J.* 61 (1990), no. 2, 445–517.

[SGA 2] Grothendieck, A.: *Cohomologie locale des faisceaux cohérents et théorèmes de Lefschetz locaux et globaux* (SGA 2). Advanced studies in pure mathematics. Masson & CIE, Paris. North-Holland, Amsterdam, 1962.

[EGA II] Grothendieck, A.: Étude globale élémentaire de quelques classes de morphismes. EGA II. *Publ. Math. IHES* 8 (1961).

[EGA IV^2] Grothendieck, A.: Étude locale des schémas et des morphismes de schémas. EGA IV^2. *Publ. Math. IHES* 24 (1965).

[EGA IV^4] Grothendieck, A.: Étude locale des schémas et des morphismes de schémas. EGA IV^4. *Publ. Math. IHES* 32 (1967).

[Hi] Hida, H.: On p-adic Hecke algebras for GL_2 over totally real fields. *Ann. of Math.* (2) 128 (1988), no. 2, 295–384.

[Jo] Jochnowitz, N.: A study of the local components of the Hecke algebra mod ℓ. *Trans. Amer. Math. Soc.* *270* (1982), no. 1, 253–267.

[Ka1] Katz, N. M.: p-adic properties of modular schemes and modular forms. In *Modular functions of one variable, III* (Proc. Internat. Summer School, Univ. Antwerp, Antwerp, 1972), pp. 69–190. Lecture Notes in Mathematics 350. Springer-Verlag, 1973.

[Ka2] Katz, N. M.: Higher congruences between modular forms. *Ann. of Math.* (2) 101 (1975), 332–367.

[Ka3] Katz, N. M.: A result on modular forms in characteristic p. In *Modular functions of one variable, V* (Proc. Second Internat. Conf., Univ. Bonn, Bonn, 1976), pp. 53–61. Lecture Notes in Math., Vol. 601, Springer, Berlin, 1977.

[Ka4] Katz, N. M.: p-adic L-functions for CM fields. *Invent. Math.* 49 (1978), no. 3, 199–297.

[La] Lang, S.: *Algebra.* Second edition. Addison-Wesley Publishing Company, Advanced Book Program, Reading, MA, 1984.

[Me] Messing, W.: *The crystals associated to Barsotti-Tate groups*, Lecture Notes in Mathematics 264, Springer-Verlag, 1972.

[MiSt] Milnor, J. W.; Stasheff, J. D.: *Characteristic classes.* Annals of Mathematics Studies, No. 76. Princeton University Press, Princeton, N. J.; University of Tokyo Press, Tokyo, 1974.

[vdP] Van der Poorten, A. J., te Riele, H. J. J., Williams, H. C.: Computer verification of the Ankeny-Artin-Chowla conjecture for all primes less than 100 000 000 000. *Math. Comp.* 70 (2001), no. 235, 1311–1328. Corrigenda and addition: *Math. Comp.* 72 (2003), no. 241, 521–523.

[Ra] Rapoport, M.: Compactifications de l'espace de modules de Hilbert-Blumenthal. *Compositio Math.* 36 (1978), no. 3, 255–335.

[Ri] Ribet, K. A.: p-adic interpolation via Hilbert modular forms. *Algebraic geometry* (Proc. Sympos. Pure Math., Vol. 29, Humboldt State Univ., Arcata, Calif., 1974), pp. 581–592. Amer. Math. Soc., Providence, R. I., 1975.

[Se] Serre, J.-P.: Formes modulaires et fonctions zêta p-adiques. *Modular functions of one variable, III* (Proc. Internat. Summer School, Univ. Antwerp, 1972), pp. 191–268. Lecture Notes in Math., Vol. 350, Springer, Berlin, 1973.

[Si] Siegel, C. L.: *Lectures on the analytic theory of quadratic forms (1934-35).* Institute for Advanced Studies Lecture Notes. Reprinted 1955.

[Wa] Washington, L. C.: *Introduction to Cyclotomic Fields.* Graduate Text in Mathematics 83. Springer-Verlag, 1982.

[Zi] Zink, T.: The display of a formal p-divisible group. *Cohomologies p-adiques et applications arithmtiques, I.* Astérisque No. 278 (2002), 127–248.

Editorial Information

To be published in the *Memoirs*, a paper must be correct, new, nontrivial, and significant. Further, it must be well written and of interest to a substantial number of mathematicians. Piecemeal results, such as an inconclusive step toward an unproved major theorem or a minor variation on a known result, are in general not acceptable for publication. Papers appearing in *Memoirs* are generally longer than those appearing in *Transactions*, which shares the same editorial committee.

As of September 30, 2004, the backlog for this journal was approximately 5 volumes. This estimate is the result of dividing the number of manuscripts for this journal in the Providence office that have not yet gone to the printer on the above date by the average number of monographs per volume over the previous twelve months, reduced by the number of volumes published in four months (the time necessary for preparing a volume for the printer). (There are 6 volumes per year, each containing at least 4 numbers.)

A Consent to Publish and Copyright Agreement is required before a paper will be published in the *Memoirs*. After a paper is accepted for publication, the Providence office will send a Consent to Publish and Copyright Agreement to all authors of the paper. By submitting a paper to the *Memoirs*, authors certify that the results have not been submitted to nor are they under consideration for publication by another journal, conference proceedings, or similar publication.

Information for Authors

Memoirs are printed from camera copy fully prepared by the author. This means that the finished book will look exactly like the copy submitted.

The paper must contain a *descriptive title* and an *abstract* that summarizes the article in language suitable for workers in the general field (algebra, analysis, etc.). The *descriptive title* should be short, but informative; useless or vague phrases such as "some remarks about" or "concerning" should be avoided. The *abstract* should be at least one complete sentence, and at most 300 words. Included with the footnotes to the paper should be the 2000 *Mathematics Subject Classification* representing the primary and secondary subjects of the article. The classifications are accessible from `www.ams.org/msc/`. The list of classifications is also available in print starting with the 1999 annual index of *Mathematical Reviews*. The Mathematics Subject Classification footnote may be followed by a list of *key words and phrases* describing the subject matter of the article and taken from it. Journal abbreviations used in bibliographies are listed in the latest *Mathematical Reviews* annual index. The series abbreviations are also accessible from `www.ams.org/publications/`. To help in preparing and verifying references, the AMS offers MR Lookup, a Reference Tool for Linking, at `www.ams.org/mrlookup/`. When the manuscript is submitted, authors should supply the editor with electronic addresses if available. These will be printed after the postal address at the end of the article.

Electronically prepared manuscripts. The AMS encourages electronically prepared manuscripts, with a strong preference for $\mathcal{A}\mathcal{M}\mathcal{S}$-LaTeX. To this end, the Society has prepared $\mathcal{A}\mathcal{M}\mathcal{S}$-LaTeX author packages for each AMS publication. Author packages include instructions for preparing electronic manuscripts, the *AMS Author Handbook*, samples, and a style file that generates the particular design specifications of that publication series. Though $\mathcal{A}\mathcal{M}\mathcal{S}$-LaTeX is the highly preferred format of TeX, author packages are also available in $\mathcal{A}\mathcal{M}\mathcal{S}$-TeX.

Authors may retrieve an author package from e-MATH starting from `www.ams.org/tex/` or via FTP to `ftp.ams.org` (login as `anonymous`, enter username as password, and type `cd pub/author-info`). The *AMS Author Handbook* and the *Instruction Manual* are available in PDF format following the author packages link from `www.ams.org/tex/`. The author package can be obtained free of charge by sending email to `pub@ams.org` (Internet) or from the Publication Division, American Mathematical Society, 201 Charles St., Providence, RI 02904, USA. When requesting an author package, please specify AMS-LaTeX or AMS-TeX, Macintosh or IBM (3.5) format, and the publication in which your paper will appear. Please be sure to include your complete mailing address.

Sending electronic files. After acceptance, the source file(s) should be sent to the Providence office (this includes any TeX source file, any graphics files, and the DVI or PostScript file).

Before sending the source file, be sure you have proofread your paper carefully. The files you send must be the EXACT files used to generate the proof copy that was accepted for publication. For all publications, authors are required to send a printed copy of their paper, which exactly matches the copy approved for publication, along with any graphics that will appear in the paper.

TeX files may be submitted by email, FTP, or on diskette. The DVI file(s) and PostScript files should be submitted only by FTP or on diskette unless they are encoded properly to submit through email. (DVI files are binary and PostScript files tend to be very large.)

Electronically prepared manuscripts can be sent via email to `pub-submit@ams.org` (Internet). The subject line of the message should include the publication code to identify it as a Memoir. TeX source files, DVI files, and PostScript files can be transferred over the Internet by FTP to the Internet node `e-math.ams.org` (130.44.1.100).

Electronic graphics. Comprehensive instructions on preparing graphics are available at `www.ams.org/jourhtml/graphics.html`. A few of the major requirements are given here.

Submit files for graphics as EPS (Encapsulated PostScript) files. This includes graphics originated via a graphics application as well as scanned photographs or other computer-generated images. If this is not possible, TIFF files are acceptable as long as they can be opened in Adobe Photoshop or Illustrator. No matter what method was used to produce the graphic, it is necessary to provide a paper copy to the AMS.

Authors using graphics packages for the creation of electronic art should also avoid the use of any lines thinner than 0.5 points in width. Many graphics packages allow the user to specify a "hairline" for a very thin line. Hairlines often look acceptable when proofed on a typical laser printer. However, when produced on a high-resolution laser imagesetter, hairlines become nearly invisible and will be lost entirely in the final printing process.

Screens should be set to values between 15% and 85%. Screens which fall outside of this range are too light or too dark to print correctly. Variations of screens within a graphic should be no less than 10%.

Inquiries. Any inquiries concerning a paper that has been accepted for publication should be sent directly to the Electronic Prepress Department, American Mathematical Society, 201 Charles St., Providence, RI 02904, USA.

Editors

This journal is designed particularly for long research papers, normally at least 80 pages in length, and groups of cognate papers in pure and applied mathematics. Papers intended for publication in the *Memoirs* should be addressed to one of the following editors. In principle the Memoirs welcomes electronic submissions, and some of the editors, those whose names appear below with an asterisk (*), have indicated that they prefer them. However, editors reserve the right to request hard copies after papers have been submitted electronically. Authors are advised to make preliminary email inquiries to editors about whether they are likely to be able to handle submissions in a particular electronic form.

Titles in This Series

820 **Vadim A. Kaimanovich and Mikhail Lyubich,** Conformal and harmonic measures on laminations associated with rational maps, 2005

819 **F. Andreatta and E. Z. Goren,** Hilbert modular forms: mod p and p-adic aspects, 2005

818 **Tom De Medts,** An algebraic structure for Moufang quadrangles, 2005

817 **Javier Fernández de Bobadilla,** Moduli spaces of polynomials in two variables, 2005

816 **Francis Clarke,** Necessary conditions in dynamic optimization, 2005

815 **Martin Bendersky and Donald M. Davis,** v_1-periodic homotopy groups of $SO(n)$, 2004

814 **Johannes Huebschmann,** Kähler spaces, nilpotent orbits, and singular reduction, 2004

813 **Jeff Groah and Blake Temple,** Shock-wave solutions of the Einstein equations with perfect fluid sources: Existence and consistency by a locally inertial Glimm scheme, 2004

812 **Richard D. Canary and Darryl McCullough,** Homotopy equivalences of 3-manifolds and deformation theory of Kleinian groups, 2004

811 **Ottmar Loos and Erhard Neher,** Locally finite root systems, 2004

810 **W. N. Everitt and L. Markus,** Infinite dimensional complex symplectic spaces, 2004

809 **J. T. Cox, D. A. Dawson, and A. Greven,** Mutually catalytic super branching random walks: Large finite systems and renormalization analysis, 2004

808 **Hagen Meltzer,** Exceptional vector bundles, tilting sheaves and tilting complexes for weighted projective lines, 2004

807 **Carlos A. Cabrelli, Christopher Heil, and Ursula M. Molter,** Self-similarity and multiwavelets in higher dimensions, 2004

806 **Spiros A. Argyros and Andreas Tolias,** Methods in the theory of hereditarily indecomposable Banach spaces, 2004

805 **Philip L. Bowers and Kenneth Stephenson,** Uniformizing dessins and Belyĭ maps via circle packing, 2004

804 **A. Yu Ol'shanskii and M. V. Sapir,** The conjugacy problem and Higman embeddings, 2004

803 **Michael Field and Matthew Nicol,** Ergodic theory of equivariant diffeomorphisms: Markov partitions and stable ergodicity, 2004

802 **Martin W. Liebeck and Gary M. Seitz,** The maximal subgroups of positive dimension in exceptional algebraic groups, 2004

801 **Fabio Ancona and Andrea Marson,** Well-posedness for general 2×2 systems of conservation law, 2004

800 **V. Poénaru and C. Tanas,** Equivariant, almost-arborescent representation of open simply-connected 3-manifolds; A finiteness result, 2004

799 **Barry Mazur and Karl Rubin,** Kolyvagin systems, 2004

798 **Benoît Mselati,** Classification and probabilistic representation of the positive solutions of a semilinear elliptic equation, 2004

797 **Ola Bratteli, Palle E. T. Jorgensen, and Vasyl' Ostrovs'kyĭ,** Representation theory and numerical AF-invariants, 2004

796 **Marc A. Rieffel,** Gromov-Hausdorff distance for quantum metric spaces/Matrix algebras converge to the sphere for quantum Gromov-Hausdorff distance, 2004

795 **Adam Nyman,** Points on quantum projectivizations, 2004

794 **Kevin K. Ferland and L. Gaunce Lewis, Jr.,** The $RO(G)$-graded equivariant ordinary homology of G-cell complexes with even-dimensional cells for $G = \mathbb{Z}/p$, 2004

793 **Jindřich Zapletal,** Descriptive set theory and definable forcing, 2004

792 **Inmaculada Baldomá and Ernest Fontich,** Exponentially small splitting of invariant manifolds of parabolic points, 2004